Walking with the senses

Perceptual techniques for walking in simulated environments

Federico Fontana, Yon Visell (Eds.)

Logos Verlag Berlin

λογος

Bibliographic information published by the Deutsche Nationalbibliothek

The Deutsche Nationalbibliothek lists this publication in the Deutsche
Nationalbibliografie; detailed bibliographic data are available
in the Internet at http://dnb.d-nb.de .

ISBN 978-3-8325-2967-3

Logos Verlag Berlin GmbH
Comeniushof, Gubener Str. 47,
D-10243 Berlin
Germany

Tel.: +49 (0)30 / 42 85 10 90
Fax: +49 (0)30 / 42 85 10 92
http://www.logos-verlag.de

Preface

The first inspiration to collectively study technologically-mediated interactions between the feet and ground, and the possibility to influence these interactions through novel human-computer interfaces, arose on while we were navigating the streets of Venice toward the railway station, after the conclusion of a research project meeting. The first ideas emerged while waiting for the "Vaporetto" water bus, as the boat stop bobbed gently under our feet on the waters of the lagoon, where we hypothesized preliminary directions for a research project on walking. Only later did we note the special position Venice occupied as it was evoked by Marcel Proust in his *Remembrance of Things Past* (*À la recherche du temps perdu*). The narrator of the book recalls a time when, while stepping into the courtyard of the Princess de Guermantes' residence in Paris, he is caught in a moment of accidental imbalance between two uneven flagstones, which joyfully evoked a similar experience he felt while walking at the baptistry of Piazza San Marco in Venice.

Of course, there was a more pragmatic impetus as well. Separate investigations at McGill University (on haptic interactive floors) and at the University of Verona (on footstep sound synthesis) spoke compellingly toward new possibilities that might emerge through such a joint collaboration. Although these technological tools were available, along with further expertise in areas including immersive virtual reality display and haptic engineering (at Université Pierre et Marie Curie, INRIA IRISA, and Aalborg University), the act of bringing these pieces together raised as many challenges as it solved. This included both pragmatic questions, such as how to effectively integrate different technologies, and more fundamental ones, related to the evaluation of novel, often multimodal perceptual effects that related to an interactive setting (human locomotion augmented with virtual multimodal stimuli) that had been seldom studied. The latter fit the goal of our investigation, which, complementing preexisting work on the design of walking interfaces for navigation in primarily visually rendered virtual environments, instead directed attention to the rich variety of multisensory phenomena that emerge when the foot and body enter into contact with a firm ground surface composed of complex, natural materials. The sounds, bodily movements, and tactile sensations we experience as a result signify the spaces we traverse in an intuitive and familiar way, and communicate to us their characteristics. The ecological information we thus acquire allows us to balance, navigate and orient during everyday tasks in unfamiliar environments, by making use of the invariant ecological meanings that we have learned through extensive prior experience with walking tasks.

The ambitious and wide-ranging research plan we embarked on in the *Natural Interactive Walking* project is reflected in the contents of the present volume. Its goals included the design, prototyping and evaluation of multimodal floor-based machine interfaces for the interactive augmentation of otherwise "neutral" (i.e., flat, silent, and visually homogeneous) ground surfaces; the study of multisensory effects and cross-modal illusions, involving the senses of touch, kinesthesia, audition, and vision, that were made possible by the novel interfaces that were explored; the realization of rich interaction contexts based on ensembles of the previous interface designs; the study of foot-floor interaction paradigms occurring in these contexts, and of the opportunities that they may offer thanks to the high sensitivity of the human feet. Applications of such paradigms may include active signalling for simplified navigation in functional spaces, support to human labor in hostile environments, and healthcare applications for aiding people with reduced foot sensitivity and during rehabilitation of impairments affecting locomotion.

To innovate within a field encompassing such a wide spectrum of knowledge requires competence and technical skills in a diverse array of topics that are not accessible to any single scientific or engineering researcher, that are essentially collaborative and interdisciplinary. However, we took encouragement from the existence of compelling research in related interdisciplinary research areas, including haptic interaction for the hands and human interaction in virtual reality environments. Indeed, several of the technologies and methodologies described in the chapters of this book can be said to be borrowed, at least in part, from those fields. Success in this endeavor could only be achieved through the collaboration of an ensemble of researchers organized around an ambitious set of interrelated goals, which also permitted sufficient independence for them to develop constituent competencies. The research plan was enriched with further components having an orientation toward applications in virtual reality: among these, the development of novel interfaces for navigation in virtual spaces.

The Natural Interactive Walking project was realized thanks to funding from the European Union and the Québec Ministère de Developpement Economique, Innovation, et Exportation which permitted a team of researchers to collaborate on the themes described above for a total of three years. In providing a survey of the most interesting results of this project, the present book represents one of the most important products of the project itself. The result is far from free of defects, as the strong interdisciplinary character of the research makes it difficult to integrate it into a global view. As a result, and in spite of the editorial efforts made in bringing together the present content, we expect that the reader will notice the discontinuities that we, as principle project investigators, have endeavored to smooth out during the course of evolution of the research.

Finally, the production of this book occurred within a necessarily highly constrained schedule, which required, among other things, that relatively fresh research contents be digested and organized quickly in order to finalize the publication prior to the completion of the project itself. While this may have diminished the quality, it has allowed us to make this book available in the public domain. Ultimately, we think this is a worthy outcome, one of its most dear qualities.

As one final note at the outset, as editors, we agreed that we would be listed in the bibliographic data in alphabetic order, since any classification based on the order of names included would not be able to acknowledge the joint effort we have both made on the book, acknowledging from the beginning that neither editor's work was more important to realizing this book.

Udine, Paris *Federico Fontana*
September 2011 *Yon Visell*

Acknowledgments

This book would have not seen the light without the activity of several persons, who have contributed to the research presented along the chapters. Among them, we acknowledge Bruno Arnaldi, Federico Avanzini, Stephane Beniak, Amir Berrezag, Roberto Bresin, Gary Chaw, Smilen Dimitrov, Carlo Drioli, Mathieu Emily, Aurelién Le Gentil, Bruno L. Giordano, Sébastien Hillaire, Jessica W. Ip, Paul G. Kry, Alvin W. Law, Stephen McAdams, Guillaume Millet, Stefano Delle Monache, Fabio Morreale, Franck Multon, Martin J.-D. Otis, Benjamin V. Peck, Julien Pettré, Pietro Polotti, Tony Regia-Corte, Davide Rocchesso, Orianne Siret, Jan M. Wiener, and Anna de Witt.

The research leading to the content of this book received funding from the Québec Ministére de Developpement Economique, Innovation, et Exportation and from the European Community's Seventh Framework Programme under FET-Open grant agreement 222107 NIW - Natural Interactive Walking. The diffusion at no cost of this book on the World Wide Web has been made possible thanks to funding from the European Community under the same grant agreement. We acknowledge Stefan Glasauer, Sile O'Modhrain, and Carsten Preusche for their suggestions made on early deliverable book material, and Paul Hearn for his constant support to a constructive project spirit which, among other things, accompanied also the production of the book.

Contents

List of Contributors

Laurent Bonnet
INRIA, Rennes, France
e-mail: `laurent.bonnet@inria.fr`

Gabriel Cirio
INSA / INRIA, Rennes, France
e-mail: `gabriel.cirio@inria.fr`

Marco Civolani*
DIMI, University of Udine, Udine, Italy
e-mail: `marco.civolani@uniud.it`

Jeremy Cooperstock
CIRMMT, McGill University, Montreal, Canada
e-mail: `jer@cim.mcgill.ca`

Mathieu Emily
Université Rennes 2, Rennes, France
e-mail: `mathieu.emily@uhb.fr`

Federico Fontana*
DIMI, University of Udine, Udine, Italy
e-mail: `federico.fontana@uniud.it`

Vincent Hayward
ISIR, Université Pierre et Marie Curie, Paris 6, France
e-mail: `vincent.hayward@isir.upmc.fr`

Anatole Lécuyer
INRIA, Rennes, France
e-mail: `anatole.lecuyer@inria.fr`

Maud Marchal
INSA / INRIA, Rennes, France

e-mail: maud.marchal@inria.fr

Rolf Nordahl
Medialogy, Aalborg University Copenhagen, Copenhagen, Denmark
e-mail: rn@media.aau.dk

Stefano Papetti*
ICST, Zurich University of the Arts, Zurich, Switzerland
e-mail: Stefano.Papetti@zhdk.ch

Rishi Rajalingham
CIRMMT, McGill University, Montreal, Canada
e-mail: rishi@cim.mcgill.ca

Stefania Serafin
Medialogy, Aalborg University Copenhagen, Copenhagen, Denmark
e-mail: sts@media.aau.dk

Severin Smith
CIRMMT, McGill University, Montreal, Canada
e-mail: sparky@cim.mcgill.ca

Léo Terziman
INSA / INRIA / DGA, Rennes, France
e-mail: leo.terziman@inria.fr

Luca Turchet
Medialogy, Aalborg University Copenhagen, Copenhagen, Denmark
e-mail: tur@media.aau.dk

Yon Visell
ISIR, Université Pierre et Marie Curie, Paris 6, France
e-mail: yon@zero-th.org

(∗): formerly at DI, University of Verona, Verona, Italy,

Acronyms

ANOVA	Analysis of variance
CPU	Control Processing Unit
FSR	Force Sensing Resistor
GPU	Graphics Processing Unit
GRF	Ground Reaction Force
HCI	Human-Computer Interaction
HMD	Head Mounted Display
IOI	Inter-Onset Interval
std	standard deviation
VE	Virtual Environment
VR	Virtual Reality

Chapter 1

Introduction: New technologies for walking with the senses

Y. Visell and F. Fontana

In a 1939 paper, J. A. Hogan analyzes the incident in which Marcel Proust entered the courtyard of the Princess de Guermantes' residence in Paris when:

> His feet came to rest on two uneven flagstones, and as he balanced from one to the other a delicious sensation swept through his body. Proust continued to exploit this sensation which brought him so much joy [...] he swayed back and forth on the two uneven flagstones, oblivious to his surroundings. But what? Where? How came he by these sensations? [...] Then suddenly it was revealed to him. It was Venice. One day, long since past, he had stood in the baptistry of St. Mark's in Venice, balanced on two uneven flagstones. His present experience balancing on the stones in the Guermantes' courtyard was sufficiently like that one in the past to call up from within him that day in Venice, which he had for so long a time kept buried deep inside him. [120]

The experience Proust felt upon walking into this courtyard is regarded as fundamental to the creation of his celebrated *Remembrance of Things Past* and to the aesthetic theory that pervades all his Recherche. Leaving the significance of this event to Proust's aesthetic universe aside, it is worth noting the impact that the activity of exploring the flagstones with his foot had in awakening pieces of his previously acquired experience. Significantly, the association he makes is strongly non-visual. It is moreover unlikely that the two flagstones he depresses in Paris are particularly similar in shape, size or configuration to those he earlier encountered at the baptistry of St. Mark's in Venice. These few observations can also be read as containing the essential seeds that this book attempts at fertilizing, from a quite different, scientific, perspective.

The activation of memory through interaction with the external world lies near the core of Proust's experience. Compared to the experiences, visual, olfactory, and gustatory, that achieved this during other incidents of his books – as he was looking through the trees of the church spires of Martinville, and, most famously, while dipping a madeleine cake in herbal tea – at least two major differences can be noted. First, the experience is strongly kinesthetic, linked to the perception and active movement of his body. Second, it is non-visual, stemming from somatosensations involving touch, equilibrium, proprioception, and perhaps even reinforced by the sounds that were produced during the acts of movement and walking. Later, Proust's sensations would be overwhelmed by the sound of traffic in Paris, causing him to open his eyes, and to become present, attending to his immediate surroundings. The interaction depicted in the passage can also be described as "natural". Proust is fed by a number of stimuli pertaining to his everyday experience. They did not need to be mediated by any training, adaptation, or cultural translation. Any part of the Venetian experience that may have had a linguistic component, for example, would not have naturally linked to the Parisian environment. The two experiences gave rise to perceptual ecologies with significant common elements, and, through this overlap, the subject established a mental window into his memory.

Finally, Proust's perception was facilitated by means of an especially sensitive human interface that has not received enough attention in prior literature on human-computer interaction: the feet. By directly exploring certain characteristics of the surface he walked on he was able to uncover the non-visual perceptual invariants that were related through his remembrance.

Many aspects of touch sensation in the feet, including its roles in the sensori-motor control of balance and locomotion over different terrain, have been studied in prior scientific literature. Considerably less is known about how the nature of the ground itself is perceived, and how its different sensory manifestations (touch, sound, visual appearance) and those of the surroundings contribute to the perception of properties of natural ground surfaces, such as their shape, irregularity, or material composition, and our movement upon them. Not surprisingly then, little research attention in human-computer interaction and virtual reality has been devoted to the simulation of multisensory aspects of walking surfaces in ways that could parallel the emerging understanding that has, in recent years, enabled more natural means of human computer interaction with the hands, via direct manipulation, grasping, tool use, and palpation of virtual objects and surfaces. These advances have been greatly supported by the improved level of understanding of human hand function and behavior, at scales ranging from the microphysiology and biomechanics of the skin to high-level perceptual processes and neural or cognitive activity involved during object manipulation.

To support the collection of similar knowledge and to identify suitable mechanisms of interaction with the feet, we launched a research effort aimed at uncovering salient roles played by multisensory, and especially non-visual, perceptual information during locomotion and other interactions with the feet, and to apply the results of this investigation to advance the state-of-the-art in human-computer interaction with the feet. The present book summarizes the results of this effort, in presenting devices, methods, and interactive techniques that aim to reproduce virtual experiences of walking on natural ground surfaces, enabled primarily through the rendering and presentation of virtual multimodal cues of ground properties, such as texture, inclination, shape, material, or other affordances, in the Gibsonian sense. As suggested by the discussion above, the results presented here are organized around the hypothesis that walking, by enabling rich interactions with floor surfaces, consistently conveys enactive information that manifests itself through multimodal cues, and especially via the haptic and auditory channels. In order to better distinguish this investigation from prior work, we adopt a perspective in which vision plays a primarily integrative role linking locomotion to obstacle avoidance, navigation, balance, and the understanding of details occurring at ground level. The ecological information we obtain from interaction with ground surfaces allows us to navigate and orient during everyday tasks in unfamiliar environments, by means of invariant meanings that we have learned through prior experience with walking tasks.

On the side of interactive technology, the realization of novel interfaces of any kind proceeds, ideally, through design and evaluation cycles, often involving several iterations and multiple layers, including basic engineering, design, and perceptual and usability assessment. For highly interactive and multimodal interfaces such as those presented in the chapters that follow, the challenge of evaluation from either of these perspectives is especially pronounced, and this is even more true when the interactions themselves are intended to have an ecological character – that is, preserving, as far as possible, the complex perception-action relationships encountered in real world environments. Due to the developmental and scientific challenges involved, the results described here constitute, in many ways, pointers to directions of departure for this research – first steps, so to speak, toward the realization of natural walking interfaces.

This book is organized in a way that reflects the heterogeneity of the topical area it addresses. The engineering of tactile interfaces for the foot represents a new, and still nascent, development in haptics, and this novelty is reflected in the contents of the chapter by Hayward et al., which introduces new actuator and display technologies for presenting vibrotactile stimuli to the foot. The two chapters by Visell et al. treat applications of computationally augmented floor surfaces for basic tasks in human-computer interaction - namely, interaction with floor-based touch surface interfaces and tracking of persons via in-floor force sensors. Chapters by Terziman et al. and Marchal et al. respectively describe methods for navigating in immersive virtual reality environments and for rendering "pseudo-haptic" effects, which are capable of modulating the perceived kinesthetic sense of slope, ground shape, or material by means of visuo-haptic or visuo-vestibular cross modal interactions. In their chapter, Serafin et al. review some of the methods available for synthesizing the sound of virtual footsteps in response to real-time data specified by either movements of an individual in a virtual environment or by another real-time control process. They also describe the use of such synthetic stimuli in experiments on the perception of walking sounds and soundscapes containing them. In contrast to these unimodal assessments of the perception of walking, the chapter by Nordahl et al. presents the results of a number of multi- or cross-modal perceptual effects related to walking in virtual environments. These may be useful for render specific effects in an immersive virtual environment (e.g., ground slope, softness) or for increasing the level of realism, immersion, or presence felt by users of such an environment. Finally, the chapter of Cirio et al. presents physically-based techniques, and a number of evocative interactive scenarios, for realizing fully multimodal, interactive experiences of walking on virtual ground surfaces.

It is hoped that this collection may prove interesting for researchers in related fields of engineering, computing, perception, and the movement sciences, and, further, that a positive result of the incomplete nature of the results presented here may lie in its ability to suggest directions for future research.

Chapter 2
Novel haptic displays for walking interactions

V. Hayward, Y. Visell, S. Serafin, F. Fontana, and M. Civolani

Abstract This chapter describes three approaches to hardware design for foot-oriented haptic devices. The first approach entails tiling a floor and actuating each individual tile according the movements of the walkers and the effects that are desired. Each tile is responsive to the forces applied by the foot. The second approach is to embed vibrotactile transducers in shoes. Interaction forces resulting from foot pressure are sensed with a view to produce virtual interactions with ground surfaces. A wireless communication architecture is described that makes it possible to eliminate the cumbersome cables that may reduce the applicability of worn devices in a virtual reality settings. A third approach is to haptically enable an existing interface. Here, this approach is exemplified by haptically enabling an exercise device known as a wobble board.

2.1 Introduction

When the shoed foot hits the ground, a series or mechanical events ensue. There can be an impact, or merely a soft landing, according to the type of shoe, the type of ground, and the stride of the walker. Once the initial transitory effects have vanished and until the foot lifts off the ground, there can be crushing, fracturing, or hardly anything at all if the ground is stiff. There can also be slipping if the ground is solid, or soil displacement if the ground is granular. There can be other mechanical effects, such as soil compacting.

The question of what form of haptic signal to reproduce in VR applications is therefore not so simple to answer. The sense of touch, which in the foot has properties comparable to that in the hand, is not easy to fool. It has, in fact, great discriminative acumen, even through a sole [111]. But like vision or audition, in accordance to the perceptual task, it may be satisfied by little. Simply think of carrying a conversation on a portable phone—which is far from providing any amount of fidelity—or watching a black and white TV which dispenses an optical field that is hardly realistic.

The success of these transducers is first and foremost linked to their ability to deliver a minimal level of temporal and spatial information. Anything above this level is enhancement. In the case of foot, our habit to wear shoes plays in our favor since shoes filter out most of the distributed aspects of the haptic interaction with the ground, save perhaps for a distinction between the front and back of the foot at the moment of the impact. In that sense, wearing a shoe is a bit like interacting with an object through a hand-tool. The later case, as is well known, is immeasurably easier to simulate in VR than direct interaction with the hand.

When it comes to stimulating the foot, the options are intrinsically limited by the environmental circumstances. While it is tempting to think of simulating the foot by the same methods as those used to stimulate the hand [113, 114], this option must be discarded in favor of approaches that are specific to the foot. In particular, options involving treadmills, robot arms and other heavy equipment will remain confined to applications where the motor aspects dominate over the perceptual aspects of interacting with a ground [137, 236, 60, 72].

The fidelity of a haptic device depends on a number of factors, including the selection and arrangement of sensing and actuating components, and on its structural mechanical design. The structural design of vibrotactile displays has, in the past, received less attention. In addition to the devices described in other sections of this chapter, the vibrotactile augmentation of touch surfaces has been widely investigated for HCI applications [243, 101, 202], although design issues affecting their perceptual transparency have often been neglected.

In the following, we describe two, non necessarily exclusive, approaches to stimulating the foot with the simulated high-frequency components, viz. 30–800 Hz, of a foot-ground interaction. As it turns out, a great deal of sensation can be obtained this way, including sensations that tradition would normally ascribe to kinesthesia [311]. Auditory feedback is generated as well by the resulting prototypes, as a by-product of the vibrotactile actuators aboard them. One approach is to tile a floor and actuate each tile according to the movement and interaction of the walker or the user. Another approach is to provide the walker with shoes augmented with appropriate transducers. Three foot-stimulating devices are now described, starting with floor-based stimulator, continuing with a shoe-based stimulator, and ending with a haptically augmented training device.

2.2 Vibrotactile floor display

The first design is based on a high fidelity vibrotactile interface integrated in a rigid surface. The main application for which this was envisioned is the vibrotactile display of virtual ground surface material properties for immersive environments. The device consists of an actuated composite plate mounted on an elastic suspension, with integrated force sensors. The structural dynamics of the device was designed to enable it to accurately reproduce vibrations felt during stepping on virtual ground materials over a wide range of frequencies. Measurements demonstrated that it is capable of reproducing forces of more than 40 N across a usable frequency band from 50 Hz to 750 Hz.

In a broader sense, potential applications of such a device include the simulation of ground textures for virtual and augmented reality simulation [313] or telepresence (e.g., for remote planetary simulation), the rendering of abstract effects or other ecological cues for rehabilitation, the presentation of tactile feedback to accompany the operation of virtual foot controls, control surfaces, or other interfaces [314], and to study human perception. In light of the latter, an effort was undertaken to ensure a high fidelity response that would avoid artifacts and enable the presentation of carefully controlled stimuli in experiments involving human vibrotactile perception.

This device represents the first systematically designed vibrotactile floor component for haptic human-computer interaction. Passive floor-based vibrotactile actuation has been used to present low frequency information in audiovisual display applications, for special effects (e.g., vehicle rumble), in immersive cinema or VR settings [297]. The fidelity requirements that must be met by an *interactive* haptic display are, all things being otherwise equal, higher, since its users are able to actively sample its response to actions of the feet.

The interface of the device (Figure 2.1) consists of a rigid plate that supplies vibrations in response to forces exerted by a user's foot, via the shoe. The total normal

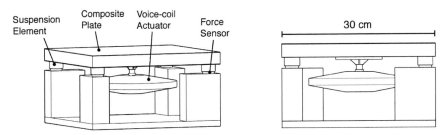

Fig. 2.1: Vibrotactile floor interface hardware for a single tile unit. Middle: View showing main components. Right: Side view with top dimension.

force $F(t)$ applied to the plate by a user is measured. It can be assumed to consist of two components: isolated transients with high frequency content, generated by foot impacts with the plate, and low-frequency forces generated by active human motions, limited in bandwidth to no more than 10 Hz [51, 309]. Vibrotactile feedback is assumed to be constrained, due to actuator limitations, to frequencies greater than a minimum value on the order of 40 Hz. A haptic simulation provides feedback approximating the vibration response felt during interaction with a virtual object. The rendering algorithms are of admittance type, computing displacements (or their time derivatives) in response to forces applied to the virtual object.

The top plate of the tile provides an interface to the body, which in the case of this device is assumed to consist of a foot wearing a shoe. Statically, the device must resist bending when loaded vertically by a force of several hundred Newtons. The rigid deflection of the plate under this load must be minimized subject to the constraint that the plate be able to vibrate freely, which can be used to infer suitable values of the mechanical parameters of the suspension elements [307].

The top plate consists of aluminum honeycomb sandwich panel component (dimensions $30.4 \times 30.4 \times 2.5$ cm, mass 400 g). This material was selected for its high bending stiffness to weight ratio. The plate is supported by cylindrical SBR rubber elastic elements positioned as shown in Figure 2.1. In dynamic or multi-tile configurations, a retaining socket surrounding the elastic support (not present in the figure) is used to keep the plate from changing position. The actuator is mounted via an aluminum bracket bonded to the center underside of the plate.

Force sensing is performed via four load cell force transducers (Measurement Systems model FX19) located below the vibration mount located under each corner of the plate. Although the cost for outfitting a single-plate device with these sensors is not prohibitive, potential applications of this device to interaction across distributed floor surface areas (see Chapter 3) may involve two dimensional $m \times n$ arrays of tiles, requiring a number $N = 4mn$ of sensors. As a result, in a second configuration, four low-cost resistive force sensors are used in place of load cells. After conditioning, the response of these sensors to an applied force is nonlinear, and varies up to 25% from part to part (according to manufacturer ratings). A measurement and subsequent linearization and force calibration of each is performed, using a calibrated load cell force sensor (details are provided in a separate publication [313]), and ensuring a linear response accurate to within 5%. Analog data from the force sensors is conditioned, amplified, and digitized via a custom acquisition board, based on an Altera FPGA, with 16-bit analog-to-digital converters. Data from each sensor is sampled at a rate of 1 kHz and transported to a host computer over UDP via the board's 10 Mbps Ethernet interface.

The tile is actuated by a single Lorentz force type inertial motor (Clark Synthesis model TST429) with a nominal impedance of 6 Ohms. The actuator has a usable bandwidth of about 25 Hz to 20 kHz, and is capable of driving the plate above strongly enough to quickly produce numbness in the region of the foot that is in contact with the tile. Digital to analog conversion of the signal driving the actuator is performed using a 24-bit, 96 kHz audio interface (Edirol model FA-101). Amplification is performed using a compact, class-D audio amplifier capable of providing 100 W to a nominal 4 Ohm actuator impedance. The resulting device is illustrated in Figure 2.2.

The magnitude frequency response was measured with a piezoelectric accelerometer (AKG model CP-411, calibrated) bonded to the top surface of the plate, as described in the caption of Figure 2.3. Frequency response measurements were taken for several different foot-plate contact conditions, while the foot was wearing a rubber soled shoe. These contact conditions modify the impedance of the display, altering its response. The results are shown in Figure 2.3.

An inverse filter H_c was further designed to equalize the device response in the frequency range from $f = 50$ Hz to 750 Hz, based on a digital IIR filter of order N, that was estimated to match the desired inverse filter using the least p-th norm opti-

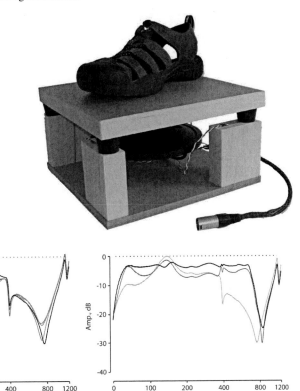

Fig. 2.2: Photo with large mens' shoe, showing representative size. The model shown is based on the low-cost force sensing resistor option. The cable in the foreground interfaces the sensors with the data acquisition unit.

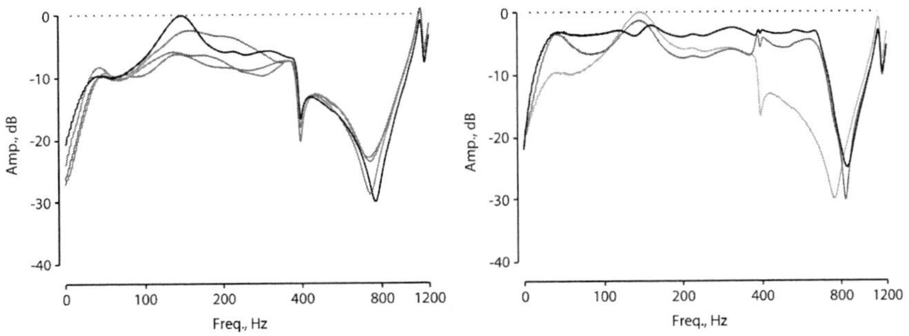

Fig. 2.3: Left: Measured log-scale magnitude frequency response of the display. Measurements were taken at a point equidistant from the plate center and edge, on a line through the center, 15 degrees from the diagonal. The free response is shown with a black line, and other foot-floor contact conditions with varying load applied via the foot are shown in gray. Right: Measured device response without foot contact, uncorrected (lightest gray) and with correction by digital IIR filters of order $N = 10$ (medium gray) and $N = 14$ (black).

mization method [291]. Figure 2.3 compares the original (free) frequency response of the device with those corrected by filters of order $N = 10$ and 14. In the latter case, the response is flat in a passband with -10 dB roll off near 50 Hz and 750 Hz.

2.3 Haptic shoes

In parallel to active floor surfaces, actuated shoes can be designed providing a similar concept. The interesting question resides on the differences in terms of types and realism of the interactions with the floor displays coming out using such two different approaches.

A pair of sandals was chosen, having soles that are easy to be customized (see Figure 2.4). Four cavities were dug in their sole, and one vibrotactile actuator (Hap-

Fig. 2.4: Haptic shoe. Left: recoil-type actuator from Tactile Labs Inc. The moving parts are protected by an aluminum shield resisting to the weight of a person. Right: approximated location of the actuators within the sandal.

tuator, Tactile Labs) was inserted in each cavity. These electromagnetic recoil-type actuators respond in the range 50–500 Hz, furthermore they can provide up to 3 g of acceleration if not too heavily loaded.

As illustrated in Figure 2.5, two actuators were bonded under the heel; the remaining two were placed under the ball of the foot. This configuration results in

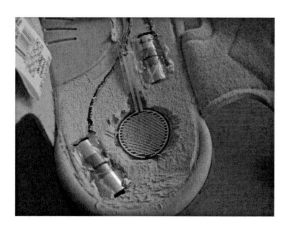

Fig. 2.5: Pressure sensor and actuators, embedded within the sole in correspondence of the heel.

two couples operating in correspondence of distinct areas, and ensures good transmission of the vibrations in the neighborhood of parts of the foot that are frequently stimulated during the act of walking.

Concerning the sensing part, each sole was equipped with two FSRs (I.E.E. SS-U-N-S-00039) placed in correspondence of the heel and the toe, respectively. The analog input from each FSR was digitized by means of an Arduino Diecimila board. Before conversion to digital, the analog signals from the FSRs need to be passed through adapted voltage dividers. Conversely, the output signals must be amplified before feeding the actuators. In our implementation each stereo amplifier received two signals from a multichannel professional audio card and then drove the two aforementioned couples of actuators, one couple per channel. The interface of a shoe with the rest of the system is shown in Figure 2.6.

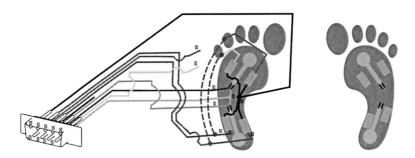

Fig. 2.6: Schematic of the port to the sensors and actuators.

The vibrotactile feedback provided by the haptuators has amplitude and bandwidth complying with the perceptual thresholds for the feet [193, 144]. However, much of the resulting perception depends on objective and subjective factors, such as type of stimulus, weight of the user, and foot position. Different coupling between the transducers and the foot, implied by perturbations in the second and third factor, are responsible for major changes in the perceived amplitude and bandwidth of the vibrotactile feedback.

In a previous prototype equipped with two TactileLabs TL002-14-A haptuators driven at high amplitudes (i.e., close to distortion level) and loaded with a normally-weighted person standing still, the sandal was measured to vibrate at the heel with the following peak amplitudes: longitudinal axis 1.5 μm; lateral axis 1.7 μm; orthogonal axis 0.9 μm. The magnitude peak in that prototype amounted to 2.4 μm. The measurements were not repeated for the new prototype. Though, similar peak amplitudes would be expected also in this case for a normally-weighted person standing still, with her feet firmly adhering to the ground.

2.3.1 Wireless implementation

This section describes the design of a wireless communication interface with the host, capable of ensuring a reliable and instantaneous tactile feedback in response to forces sensed underfoot. These features allowed to test algorithms for simulating ground surfaces, as described in subsequent chapters.

The sandals on which this architecture was tested are similar to those described above; they have Interlink model 402 FSRs fixed under the insole: one at the toe and one at the heel. The sensors were connected to four analog inputs of an acquisition board (Arduino model Duemilanove) via a voltage divider. The force signals were digitally converted using a polling procedure. For every channel, each sample was encoded in 10 bits and encapsulated within two bytes [63].

As demonstrated in previous work, despite its low cost, the Arduino Duemilanove board could function as an adequately high performance front-end for signal acquisition on personal computers through appropriate customization of the firmware [63]. By using native AVR instructions with an optimized interrupt policy, efficient operation of the micro controller's onboard ADC can be ensured. Up to six analog channels can be reliably acquired at constant sampling rate with 10 bits resolution each. Moreover, the serial-over-USB interface could be forced to send synchronous data continuously, thus obtaining uniform sampling also at the host application side.

In the wireless data acquisition setup, packets are sent from the microcontroller board through serial communication to a 2.4 GHz wireless transceiver module (http://www.sparkfun.com/products/152) based on the Nordic Semiconductor nRF2401A chip. At the other end, an identical nRF2401A module received the data stream, providing fast and reliable paired wireless communication. The receiver forwarded the data to a second micro controller board, which in turns communicated with a personal computer via USB connection. The setup is depicted in Figure 2.7.

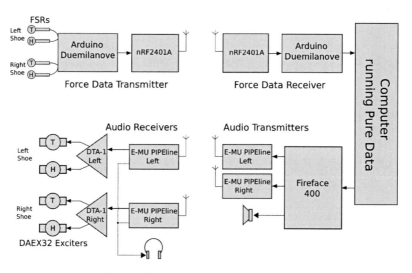

Fig. 2.7: Wireless communication architecture.

This setup was tested using a computer running an instance of the Puredata real-time environment which fetches and processes the stream, synthesizing the audio-tactile signals. Since each shoe embedded two audio-tactile transducers, the software synthesizer provided four discrete output channels: a multichannel audio interface was thus required (RME Fireface 400). This way, the stereo signal originally generated for each shoe could be mixed to mono and routed into a two-channels bus.

This setup allowed to use a stereo audio interface in the same manner as the internal chips available on ordinary laptop computers.

The four audio channels were grouped in two stereo pairs, one for each shoe, and were also transmitted wirelessly. Each pair was forwarded to an E-MU PIPEline wireless transceiver. Each transceiver was paired with an identical one working at the receiver side. Overall, a four channel audio simplex wireless connection was obtained. The line output of each E-MU at the receiver end fed a 15 W Dayton Audio DTA-1 Class D digital amplifier, driving two Dayton Audio DAEX32 transducers fixed under each sandal, one at the toe and one at the heel, closing the interaction loop. The transducers are visible in Figure 2.8, embedded in the sole.

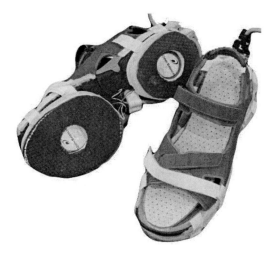

Fig. 2.8: Instrumented shoes
(bottom and top view).

Measurements performed on a previous prototype using the same data acquisition procedure [226] showed that the sensor sampling rate per analog channel depends on the speed of the serial communication and on the number of channels: with 115,200 bps, a sampling rate $F_s = 5882$/channels Hz was obtained, which corresponds to approximatively 1470 Hz per channel using four sensors. After properly adjusting the audio buffer size, the input-output latency of the entire system including the synthesis, (i.e., from the acquisition board analog inputs to the analog outputs of the audio card) amounted to 15 ms.

2.4 Haptic wobble board

The wobble board, see Figure 2.9, is a sport training tool that is primarily intended to strengthen one's ankles, improve sense of equilibrium, and work as a support to rehabilitation [183]. The wobble board is commonly made of a wooden disk covered with a rubber coat. A half-spherical stand makes it unstable: standing on

Fig. 2.9: Instrumented wobble board.

this devices requires good postural skill, like for instance when surfing. The involved sensory cues of equilibrium are provided by proprioceptive, vestibular and visual perception. Skilled users are able to correctly balance on the wobble board with their eyes closed.

Once instrumented, the wobble board can react to movement in a variety of ways, involving and enriching the perceptual modalities. For instance, the board can be programmed to respond visually, acoustically and vibromechanically, also providing hyper realistic feedback. As an example, users during normal practicing can watch images on a screen, listen to sounds, and feel vibrations that are all interactively synthesized.

Figure 2.9 shows a wobble board, instrumented with FSR's and vibrotactile actuators. A couple of haptuators was inserted between two wooden circular layers. The normal coating was replaced by specific rubber, providing optimal compliance with the feet also in the lateral direction. The system includes a 2-D accelerometer (from Phidgets, Inc.) mounted between the two layers, which measures tilt. This information was used to control audio, visual and haptic synthesis algorithms—see Chapters 7 and 5.

2.5 Conclusion

We have described three approaches to the design of devices enabling the simulation of foot-ground interaction, each having advantages and shortcomings. The floor-based approach is versatile, but more expensive owing to the increased amount of hardware that must be deployed. It is primarily appropriate for fixed installations that are typical of VR settings. The shoe-based approach has a fixed cost per user, is straightforward to implement, and the problem of cables can be solved through wireless communication architectures. Finally, many devices with which people interact through the foot can be "hapticized". Here, we described how a training wobble board can be "haptically enabled".

Chapter 3
Distributed human-computer interaction with augmented floor surfaces

Y. Visell, S. Smith, and J. Cooperstock

Abstract This chapter presents techniques for interactive via the feet with touch sensitive floor surfaces that are augmented with multimodal feedback. After a discussion of prior research in this area, it reviews one approach from the work of the authors, based on an array of instrumented floor tiles distributed over an area of several square meters, capable of sensing forces and rendering visual, vibrotactile and acoustic feedback. Aspects of the basic usability of such displays for human-computer interaction are also discussed, and we present the results of a preliminary usability evaluation with a target selection task, which was conducted to provide preliminary indications of the appropriate size and layout for foot-operated controls rendered through the floor tile interface. Potential applications for such interfaces are also reviewed, including a novel interactive scenario involving the navigation of immersively presented geospatial data navigated via the feet.

3.1 Introduction

To date, there has been limited research on foot-based interaction for computationally augmented environments. Arguably, one reason for this has been a lack of efficient interfaces and interaction techniques capable of capturing touch via the feet over a distributed display. In the present chapter, we describe techniques for interacting with computationally augmented floor surfaces, including the design of an interface based on a distributed network of low-cost, rigid floor tile components, with integrated sensing and actuation. Taking advantage of the particular structure of this interface, we draw on contact sensing techniques, through which we capture foot-floor contact loci with finer resolution than would be achieved if a single tile were regarded as the smallest relevant spatial unit.

The techniques described here may also be used to mediate interactions between a walker and a virtual ground surface. Several interactive scenarios are described in

Chapter 8, with users enabled to walk on a virtual frozen pond or along a virtual seaside beach.

3.2 Background

3.2.1 Application domains for foot-floor human computer interaction

Foot-based human-computer interaction is gaining interest as a new means of interacting in virtual reality or in ambient computing spaces, with potential applications in areas such as architectural visualization, immersive mission training, or entertainment. Although many of these applications may be amenable to interaction within traditional mouse-click and scrolling or finger-based multitouch paradigms, we believe that manipulation and parsing of the immersive data sets involved could benefit from novel interaction paradigms that leverage the naturalness of interaction on foot and the greater number of degrees of freedom possible via sensing of body movements.

Virtual floor controls could be advantageous in areas of man-machine interaction in which foot operated controls or interfaces are already commonplace, such as manufacturing assembly and repair, mass transportation, vehicle operation, or dentistry. A virtualized display can be used to provide access to instrumentation or machine controls in a way that is flexible to different contexts of use (e.g., different dental procedures). Such a virtualized display can be used to provide access to instrumentation or machine controls in a way that is flexible to different contexts of use (e.g., different dental procedures). This approach may, for example, be able to ameliorate acknowledged problems with the proliferation of physical foot pedals and other controllers in medical interactions [323].

Other applications of floor-based interfaces, as enhancements to pedestrian navigation or map-based visualization in immersive environments could emerge as particularly appropriate, insofar as they may integrate the role of the foot in self-motion. Previously investigated applications of immersive virtual reality, such as architectural walkthroughs or training simulations, may benefit from the addition of context-based interactive maps or menus, as, for example, in the investigation of LaViola et al. [158]. Other relevant application fields could include entertainment, music performance, gaming, or advertising, where companies such as Gesturetek (Fig. 3.1) and Reactrix have successfully commercialized interactive, floor-based visual displays for marketing purposes.

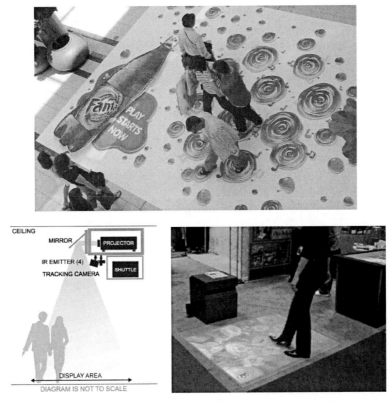

Fig. 3.1: The Ground FX video-based interactive floor display system offered by the company Gesturetek can be used for purposes such as entertainment or product promotion.

3.2.2 Techniques and technologies

Examples of the use of foot-controlled input in HCI, interactive arts and video gaming date at least to the early 1980s, with Amiga's Joyboard (1982) being one widely known example [272]. Building on the success of earlier video game control interfaces for the feet, the current Nintendo Wii balance board controller has achieved international commercial success. Research in human-computer interaction has investigated related issues in its own way, arguably beginning in the mid 1980s, when Pearson and Weiser investigated foot input devices for desktop PCs, and invented a pedal-like device called the Mole [232]. However, despite the high-level of ongoing interest in touch screen based interactions for the hands, less research and development has targeted touch-sensitive interfaces for the feet.

Direct tactile sensing for interaction with floor surfaces has conventionally been accomplished with surface-mounted force sensing arrays (e.g., [227, 266]). Such interfaces have been applied to applications including person tracking, activity tracking, or musical performance. Floor-mounted tactile sensing arrays are commercially available, but costs are high and support for real-time interaction is limited, since the predominant application areas involve offline gait and posture measurement.

Another method that has been used to enable foot-floor interactions with visual feedback employs under-floor cameras and projectors to render interactions via translucent floor plates. Gronboek et al. implemented a interactive floor surface called the iGameFloor using bodily gestures sensed optically through translucent floor plates, which were also used to display video via rear-projection. Four video projectors and cameras were installed in a cavity beneath floor-level, creating an interactive floor area 3 × 4 m in size. The locations of limbs near to the interactive surface were tracked, and were used to mediate interactions with multi-person floor-based video games. Spatialized auditory feedback was supplied by a set of loudspeakers positioned nearby.

Augsten et al. adapted the method of optical sensing via frustrated total internal reflection to enable the dynamic capture of foot-floor contact areas through back-projected translucent floor plates [19]. This method provides direct imaging of contact area, although it does not directly reveal forces. The main drawback to this approach is that it requires the installation of cameras and projectors within a potentially large recessed space that must be available beneath the floor. The authors implemented and evaluated the usability of foot-floor touch surface interfaces using this approach (Fig. 3.2). Similar to the methods described in section 3.4, they identified salient selection points within the foot-floor contact interface in order to enable precision pointing to targets or performance of other actions through a floor-based touch screen interface.

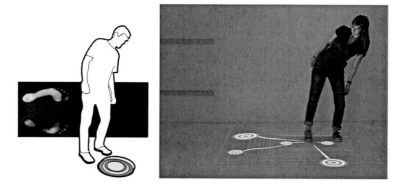

Fig. 3.2: The Multitoe interface of Augsten et al. [19] is based on the optical sensing of foot floor contact regions captured through back-projected translucent floor plates (image reproduced from [19]).

Sensations accompanying walking on natural ground surfaces in real world environments (sand in the desert, or snow in winter) are rich, multimodal and highly evocative of the settings in which they occur [309]. However, floor-based multimodal (visual, auditory, tactile) information displays have only recently begun to be investigated [312]. Related research on virtual and augmented reality environments has focused on the problem of natural navigation in virtual reality environments. Solutions such as walking in place [292] and redirected walking techniques [252] map body movements sensed through kinematic tracking onto a user's coordinates in a virtual environment (VE). The shoe-based Step WIM interface of LaViola et al. [158] introduced additional foot gestures for controlling navigation in a larger VE via a floor map, but required special shoes and did not provide auditory or haptic feedback. A number of haptic interfaces for enabling omnidirectional in-place locomotion in VEs have been developed [121], but known solutions either limit freedom in walking, or are highly complex and costly.

Fig. 3.3: Left: The floor interface is situated within an immersive, rear projected virtual environment simulator. Right: Visual feedback is provided by top-down video projection (in the instance shown, this corresponds to a virtual, multimodal sand scenario — see Chapter 8).

3.3 A distributed multimodal floor tile interface

The interface developed by the authors [314] consists of an array of rigid tiles, each of which is instrumented with force sensors (four per tile) and a vibrotactile actuator. The prototype shown in Figure 3.4 comprises a square array of 36 tiles, with an area of approximately four square meters. The floor is coated in reflective projection paint, and a pair of overhead video projectors is used for visual display, with redundant projection making it possible to reduce the effect of shadows cast by users. The individual tile interfaces are rigid, plywood plates with dimensions 30.5 × 30.5 × 2 cm, supported by elastic vibration mounts, and coupled to a vibrotactile actuator (Clark Synthesis, model TST229) beneath each plate [307]. Actuator signals are generated on personal computers, output via digital audio interfaces, and amplified.

Normal forces are sensed at locations below the corner vibration supports of each tile using a total of four resistive force sensors (Interlink model 402 FSR). Analog data from the force sensors is conditioned, amplified, and digitized via a set of 32-channel, 16-bit data acquisition boards based on an Altera FPGA. Each sensor is

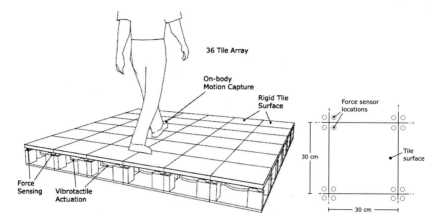

Fig. 3.4: Left: Diagrammatic view of the interface. Sensing and actuating components are integrated beneath the floor. Right: View from above showing sensor locations.

sampled at a rate of up to 1 kHz, and the data is transmitted in aggregate over a low-latency Ethernet link. An array of six small form factor computers is used for force data processing and audio and vibrotactile rendering. A separate, networked server is responsible for rendering visual feedback and managing user input.

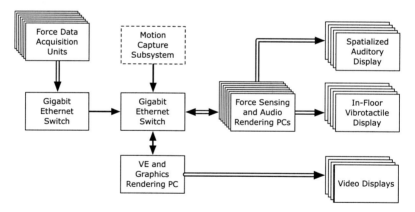

Fig. 3.5: Overview of system components and logical connections between them.

3.4 Contact sensing

For processing sensor data, we draw on intrinsic contact sensing, which aim to resolve the locations of contact, the forces at the interface, and the moment about the contact normals using internal force and torque measurements [36]. It is assumed to involve contact between a rigid object and a soft body (here, a foot), and has been previously used for robotic manipulation. Here we use it for capturing foot-ground interactions via in-floor force sensors. It requires a number of sensors on the order of the number of rigid degrees of freedom of the structure, far fewer than are needed for tactile sensing via surface mounted arrays. Since a tile has fewer sensors than rigid body degrees of freedom, we make the simplifying assumption of frictionless contact via a normally directed pressure distribution, and further assume that the relative displacement of the suspension elements in the tile is negligible. The problem is then to resolve the location of a contact centroid $\mathbf{x}_c$ associated with a normal force distribution $p_R(\mathbf{x})$ within an area R. In our context, $\mathbf{x}_c$ is a contact point such that a normal force F_c at $\mathbf{x}_c$ gives rise to the same measurements as $p_R(\mathbf{x})$ does [36]. For a floor tile with sensor locations $\mathbf{x}_j$ where measurements f_j are taken (j indexes the tile sensors), $\mathbf{x}_c$ and the normal force $\mathbf{F}_c = (0,0,F_c)$ can be recovered from scalar measurements $\mathbf{F}_j = (0,0,f_j)$ via the force and torque equilibrium equations,

$$\sum_{j=1}^{4} f_j + F_c + f_p = 0 \tag{3.1}$$

$$\sum_{j=1}^{4} \mathbf{x}_j \times \mathbf{F}_j + \mathbf{x}_c \times \mathbf{F}_c + \mathbf{x}_p \times \mathbf{F}_p = 0. \tag{3.2}$$

$\mathbf{F}_p = (0,0,f_p)$ is the weight of the the plate and actuator at the tile's center $\mathbf{x}_p$. The three nontrivial scalar equalities (3.1, 3.2) yield:

$$F_c = \sum_{i=1}^{4} f_i - f_p, \quad \mathbf{x}_c = \frac{1}{F_c} \left(\sum_{i=1}^{4} (\mathbf{x}_i - \mathbf{x}_p) f_i + f_c \mathbf{x}_p \right) \tag{3.3}$$

The contact centroid lies within the convex hull of the contact area (dashed line, Figure 3.6) at the centroid of the pressure distribution [36], and thus provides a concise summary of the foot-floor contact locus, but not about shape or orientation. When the foot-floor contact area R overlaps multiple tiles, a pressure centroid $\mathbf{x}_c$ for the entire area can be computed from those $\mathbf{x}_{ck}$ for each tile (computed from Eq. (3.3)), via $\mathbf{x}_c = w_1 \mathbf{x}_{c1} + w_2 \mathbf{x}_{c2}$, where $w_k = F_i/F$. The domain-independence of this result makes it possible to continuously track contact across tile boundaries. The difference vector $\delta \mathbf{x} = \mathbf{x}_{c1} - \mathbf{x}_{c2}$ provides additional shape information about the orientation of the contact distribution at the boundary. For a linear pressure distribution spanning the two tiles, the direction vector $\mathbf{n} = \delta \mathbf{x}/|\delta \mathbf{x}|$ is the orientation, and $|\delta \mathbf{x}|$ is half of the length of R. For convex contact shapes that are less sharply oriented (such as those of a foot), the range of angles is compressed around the edge normal direction.

Figure 3.7 shows the sequence of contact centroid locations produced by an individual walking across the floor. When there is multi-tile foot-floor contact, as illustrated here, we use a simple clustering algorithm to associate nearby contact centroids that are assumed to belong to the same foot.

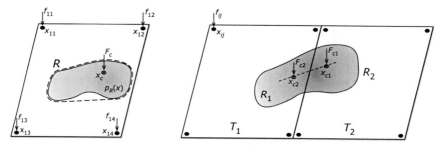

Fig. 3.6: Left: A normal force distribution $p_R(\mathbf{x})$ and associated contact centroid position $\mathbf{x}_c$. Right: A pressure distribution $p_R(\mathbf{x})$ on a region R spanning adjacent tiles. The weighted sum of centroids $\mathbf{x}_c$ is the centroid location for the distribution with support $R = R_1 \bigcup R_2$. It lies on the line segment connecting $\mathbf{x}_{c1}$ and x_{c2}. The difference $\delta\mathbf{x} = \mathbf{x}_{c1} - \mathbf{x}_{c2}$ provides information about contact shape

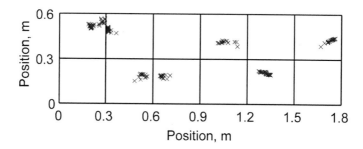

Fig. 3.7: A recorded sequence of contact centroids produced by an individual walking across the floor. Data was sampled each 100 ms to produce the figure. Each square corresponds to one floor tile. When the foot lies on a single tile, as weight shifts from heel to toe, an array of centroids is produced, moving in the direction of travel. At inter-tile boundaries, at each instant one centroid is produced on each tile that there is contact with.

Figure 3.8 presents measurements of 50 estimated contact positions determined by the method of Eq. (3.3), using a single calibrated floor tile. Contacts were measured to be localized with a typical accuracy of 2 cm, with worst-case errors of ≈ 3 cm, smaller than the linear dimensions of the tile (30 cm) or the typical width of an adult shoe. Distortion was observed to be highest, and accuracy lowest, near the edges of the tile.

3.5 Floor touch surface interfaces

The sensing methods presented above can furthermore be employed to implement virtual floor-based touch interfaces. One set of examples we have created consists of an array of standard UI widgets to be controlled with the feet (Figure 3.9). Input is based on a multi-touch screen metaphor mediated by a set of interaction points

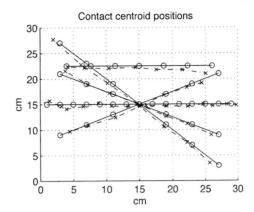

Fig. 3.8: Results of 50 measurements comparing true
normal force positions (circles) with contact centroid
estimates (Xs).

(cursor locations), which are defined as the contact centroids x_c with the largest forces. Force thresholds associated with a control are used to determine selection. The controls provide positive tactile feedback supplied by the actuators, in the form of synthesized click-like transient vibrations or sliding (friction) vibrations.

3.5.1 Interface design toolkit

We developed a software layer and network protocol to facilitate the design of interactive applications using the floor surface. It abstracts the hardware systems, which are accessed over a local Ethernet network, and connects them to the user interface. The software layer also processes the sensor data to extract foot-floor contact points that are used for interaction, and provides them with unique IDs that persist while contact is sustained. Additionally, it allows to remotely cue and present localized vibrotactile feedback in response to activation or control of user interface objects on the floor. The design of this software layer is based, in part, on the TUIO protocol for table-top touch interfaces [143]. Figure 3.9 illustrates a virtual floor-based touch interface, consisting of sliders, buttons, toggle switches, and similar elements. The controls can be operated without consideration for the location of the tile boundaries, since we track interaction points continuously across tile boundaries. Normal force thresholds are used to determine when buttons or other controls are being engaged. Concurrent audio and vibrotactile feedback, in the form of clicks, taps, or rubbing sounds or vibrations, can be assigned to be supplied in tandem with the discrete or continuous response of a control.

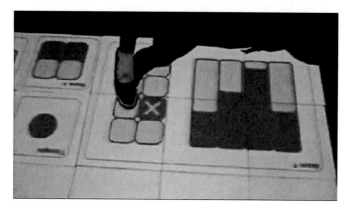

Fig. 3.9: A user interacting with floor-based interface widgets implemented with the interface design toolkit described here.

3.5.2 Usability of foot-floor touch surface interfaces

Due to the novelty of floor-based touch surface interfaces, a question that is soon encountered when beginning to design applications for them concerns the appropriate size and layout of virtual controls, and related features, such as informational overlays or menu structures. As in manual touch surface interaction, the appropriate size likely depends on a range of factors, including sensing limitations, users' motor abilities, target parameters, and feedback modalities, as has been extensively studied and modeled in the human-computer interaction literature [178, 140] and the subset of literature on touch screen usability [34, 9]. The size appropriate for touch screen controls has been shown to depend on the interaction technique adopted. Precision control strategies can enable single pixel accuracy in finger-based touch screen interaction [9, 242], and related techniques may prove effective for use with the feet. Limited research to date has addressed the usability of floor based touch interfaces. However, since the feet play very different roles in human movement than the hands do, the extent to which the kinds of techniques that have been adopted for use with the fingers may be useful for interaction via the feet is questionable.

Human movement research has investigated foot movement control in diverse settings. Visually guided targeting with the foot has been found to be effectively modeled by a similar version of Fitts' law as is employed for modeling hand movements, with an execution time about twice as long for a similar hand movement [119]. However, for many interfaces, usability is manifestly co-determined by both operator and device limitations (e.g., sensor noise or inaccuracy), providing a window on both.

Augsten et al. enabled users to select keys on an on-screen keyboard projected on the floor using precision optical motion capture tracking of the foot location and a crosshair display to indicate the location that was being pointed to, and a pressure threshold to determine pressing. They that found users were able to select target buttons (keyboard keys) having one of the three dimensions 1.1×1.7, 3.1×3.5, or 5.3×5.8 cm with respective error rates of 28.6%, 9.5%, or 3.0%. Sensing accuracy was not a factor in this study, since the motion capture input provided sub-millimeter accuracy. The authors undertook further studies to determine the optimal part of the foot to be used for selection, and to determine the extent to which non-intentional stepping actions could be discriminated from volitional selection operations.

Visell et al. investigated users' abilities to select on-screen virtual buttons of different sizes presented at different distances and directions using the interface described in the preceding section [314]. In this task, the limited resolution of the force sensors (Figure 3.8) was one factor that could influence performance. Users could activate a button by pressing it in a way that resulted in a contact centroid within the area of the button exceeding a force threshold of about 35 N. The buttons were round, ranged in diameter from 4.5 to 16.5 cm, and were presented at four distances, on lines radiating from between their feet, oriented at one of two angles. Upon selection, the buttons provided visual feedback in the form of a 20 cm white disc centered in place of the original appearance. All buttons provided the same feedback. Only the buttons and foot locations were displayed. No audio or vibrotactile feedback was provided. Summaries of the success frequencies are presented in Figure 3.10.

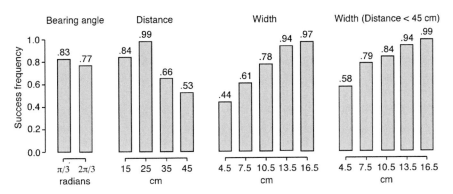

Fig. 3.10: Successful target selection rate vs. distance, angle of presentation (measured away from preferred foot), and button width, inclusive and exclusive of the farthest targets.

Users were determined to have selected larger targets within the allotted two-second interval at a significantly higher rate of success than smaller ones. Performance with the largest was very high (98%), and that for the smallest was low (44%). Small targets pose two potential problems. First, they can be occluded by the foot during selection. This problem was somewhat mitigated because when this occurred, they were often visible projected on top of the foot. Second, limitations on precise control can arise from factors such as shoe width, human motor abilities, and sensor positioning errors. Six out of eight participants in this study reported finding a strategy to activate the small buttons by using a feature of the shoe or changing the applied force. Nearby targets (distances of $D = 15$ to 25 cm) were selected at a higher rate. However, performance was better at 25 cm than at the nearest distance of 15 cm (98.5% vs. 84%). One apparent reason was that when an interface element was too close, it could be occluded from view by the body, or could present a difficult viewing angle. Due to such effects selection time T might not be expected to to follow a Fitts' Law relation, $T = a + b \log_2(D/W)$, but this was not tested here.

Although a mobile user might be able to avoid visibility problems, they seem to be an important consideration. For our device, position sensing is most accurate near the centers of the tiles, as indicated in the preceding section. This was noticed by users of the system, two of whom volunteered that they had learned to better activate small buttons that were close to edges by pressing them off-center. Neck fatigue was a frequently cited source of discomfort, suggesting that displacing the visual feedback relative to the foot-based interaction point might be beneficial.

Further work in this area is clearly needed in order to characterize the usability aspects of floor based touch screen displays. A greater understanding of factors such as control element size, display scale, motor abilities, modalities, and other aspects salient to the use of such a device will certainly be needed. In addition, it would be valuable to know to what extent usability might be improved through the use of auditory or vibrotactile feedback. Although Augsten et al. were able to identify some strategies that could be used to avoid selecting controls that are walked across, there remain open questions concerning concerns the interplay between users' movements on foot and their interactions with the touch surface. A novel aspect of interacting on foot is that, implicitly, both feet are involved in touching the interface, due to requirements of movement and of maintaining balance. In everyday actions, like striking a soccer ball, weight is often shifted onto one foot, which specifies an anchored location, while the opposite is used to perform an action. Thus, floor interfaces that involve movement may share similarities to bimanual interaction in HCI, a connection that provides another potential avenue of future investigation.

3.6 Application to geospatial data navigation

The ubiquity of multi-touch mobile devices and computers has popularized the use of two-finger gestures to navigate and control zoom level in the display of data via a touch screen. Such an approach focuses on the fingers as a means of providing input and may not be appropriate for settings in which the display surface is not amenable to finger-based interaction, when the hands are occupied, or when the data visualization application occupies a larger volume of space. We argue that this focus has limited the possibilities for carrying out complex exploration tasks in ways that leverage the capabilities we exploit naturally in the physical world.

For target acquisition tasks, e.g., menu and object selection, the accuracy and speed of the input primitives are important. In contrast, for spatial navigation tasks, e.g., panning and zooming, using dragging, resizing, and scrolling operations, less accuracy is often acceptable. Consequently, spatial navigation is likely an appropriate candidate for foot-based input, leaving the hands free, in parallel, to specialize on the more time- and accuracy-critical operations [221]. For instance, in the context of collaborative design or decision making, e.g., urban and architectural planning, or emergency and crisis management [250, 6, 177], different roles are appropriate for foot input and hand input. Specifically, the feet could be used to specify approximate location while the hands perform other more critical or complex input tasks such as target selection, annotation, or drawing in the virtual design space, or dispatching tasks in emergency situations. Moreover, this represents a compelling alternative for scenarios where use of the hands is inconvenient. For example, in the context of airport information kiosks, where users are often carrying luggage, it may be desirable and appropriate to obtain directions to one's gate through a foot-based interface.

3.6.1 Foot-based gestures for geospatial navigation

Using the distributed floor tile interface described above, we implemented an application and several foot-based interaction methods for navigating geospatial data sets presented through a multitouch-sensitive floor surface in an immersive virtual reality environment. In contrast to prior work, focused on foot-based interaction at a fixed location, our work explores the benefit of such interaction in a distributed floorspace. This approach can be seen as an alternative to desktop computer interfaces for navigating geospatial data, and may be particularly suitable for situations in which an immersive display is involved or in which users' hands are occupied with other tasks.

We developed these interactive techniques for exploring geospatial information by navigating with a floor-based map interface, presented via the multitouch floor surface. The application was developed with the user interface framework presented above. As participants navigated to specified locations using the map, via one of the foot-floor interaction techniques, the images of the streets for the locations visited were presented in real time via the wall surfaces whenever applicable. The available data was acquired from existing internet-based mapping applications from Google and Microsoft.

The interaction techniques investigated included a "pivot" interface, which uses relative foot position as a navigational input, a "magic tape" interface that uses absolute foot position within the workspace, a "sliding" interface that allowed users to virtually push themselves within the mapping interface, and a "classic" approach using virtual buttons and sliders to provide a comparison with a more conventional user interface paradigm. Additional gestures allowed participants to control the zoom level of the map.

Classic interface

The "classic" interface (Figure 3.11) transposes the basic design used for spatial navigation in mouse-based applications to the setting of floor-based interaction.

Four buttons in a cross arrangement control position and a discrete-valued slider provides control over the zoom level. The discrete slider levels match those used for the whole body gestures in the subsequent interfaces.

Fig. 3.11: Four button arrows (left) corresponding to panning in the cardinal directions and a slider (right) for zoom.

Pivot interface

The users establishes a pivot point by standing still for a short period. Placing one foot outside of the pivot area, indicated by a circle around the feet, pans in the direction specified as the vector from the pivot center to the outside foot. The participant can, at any time, exit the pivot area and establish a new one elsewhere (Figure 3.12).

Sliding interface

As with the bezel interface, the users first establishes a pivot point by standing still for a short period. Then, by placing one foot outside the pivot area and using sliding or dragging gestures, akin to touch-screen scroll on an iPhone, the user can pan the map (Figure 3.13).

Fig. 3.12: The first foot to remain still establishes the pivot point, surrounded by a blue circle. The second foot then specifies the vector, relative to the pivot, for panning.

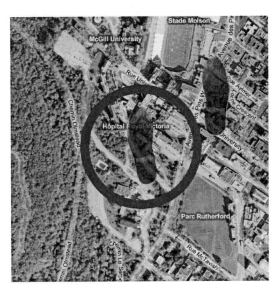

Fig. 3.13: In the sliding interface, the first foot establishes a pivot point, and the second foot slides or drags the display in a scrolling motion, with inertia.

"Magic tape" interface

The "magic tape" interface (Figure 3.14), inspired by the work of Cirio et al. (see Chapter 5), takes advantage of the larger floor surface by employing an interaction paradigm based on absolute foot position. This metaphor allows users to navigate freely in the center of the floor space, without altering the displayed map contents. However, when participants walk past the boundary region of the floor surface, the map pans in the direction designated by the user's position. The farther the user from the center, the greater the panning speed. The rate of panning is determined by a quadratic function of the magnitude of the vector from the center to the user's position.

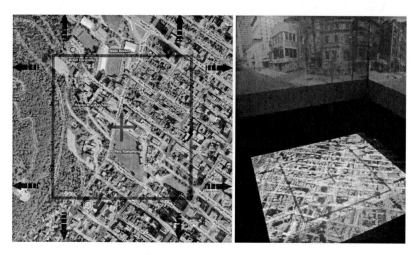

Fig. 3.14: The "magic tape" interface: A rectangular outline indicates the magic tape boundary, beyond which, the user's footstep will result in panning in the direction formed by a vector from the center crosshair and the user's foot.

3.6.2 Gesture recognition: Crouch and zooming

Unlike the direct-manipulation inspired control actions described above, gestures are recognized to allow temporally extended body movements, such as jumping, to affect the application state in a way that is not directly or instantaneously linked to changes in the user interface.

In the geospatial data navigation application, short body gestures can be used in conjunction with the control actions described above to control zooming. A curt "crouching" gesture zooms the map in, while a curt "jumping" gesture (raised onto the toes) zooms out.

Since such gestures have distinctive normal-force profiles trough time they can be matched to a reference signal. The use of a dynamic time warping algorithm for the matching in conjunction with preprocessing normalization allows for variations in gesture speed and amplitude from the user.

The analysis of these gestures, which relate to shifts in the weight and posture of users, is based on recognizing distinct sequences of whole body movements based on dynamic force signals sensed as a result of the weight shifts and posture changes. Many such movements are characterized by distinctive transient force signal from the floor sensors, due to the reaction forces the user generates when moving. A larger

Fig. 3.15: Crouching and jumping gestures for zoom control.

gestural vocabulary based on such movements could include crouching, jumping, leaning and tapping. Such movements are accompanied by force impulses as the user lifts and moves his or her body weight.

There are two challenges to detecting such gestures. First, they can vary in amplitude due to the manner and intensity of execution, and to the weight of the user. This issue can be resolved by normalizing the data. Secondly, the timing of sub-steps comprising such movements may vary between users, resulting in signals that appear distorted in their progression in time. For instance, one user's "curt jump" might consist of a quick rise to the toes followed immediately by a drop, while another user may pause briefly at the peak the rise, stand on tiptoe, and then slowly drop down. While both inputs will share a common distinctive shape, they are stretched and distorted in time relative to each other.

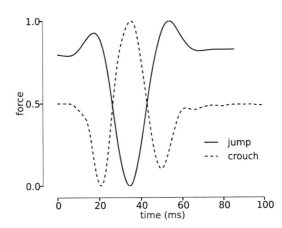

Fig. 3.16: Characteristic normal force profile of the jump and crouch gestures.

In order to compensate for this, we use standard techniques from time series recognition. A dynamic time warping algorithm (DTW) is used to match reference gestures to the input data series. The system performs a continuous windowed DTW analysis of the force input from the sensor with respect to a number of reference force waveforms, generated beforehand from a training data set. Given an input data series $i(t)$ and a reference data series $r(t)$, the DTW algorithm builds a 2D graph such that every point $p(t,y) = |i(t) - r(y)|$, and seeks to optimize the temporal alignment between the two signals, by assigning the value of the mapping function $y(t)$. The sum of all the values of p along any monotonic and continuous path (moving forward in time only) provides the matching cost of the given path, with lower costs representing better matches.

Since the "start" of any gesture can be difficult to determine prior to analysis, we relax the typical DTW constraint which requires proceeding through each sample of the input in its entirely, and instal allow the matching algorithm to enter anywhere along the bottom edge $(p(x_a,0))$ and to exit anywhere along the top $(p(x_b,y_{max}),x_b > x_a)$, which accommodates cases in which the input gesture may have been performed much faster than the reference. Once the path of least cost through the graph is computed, the similarity of the reference and input data series are obtained with allowances for time warping, allowing users flexibility to vary the speed of different segments of the gesture, as long as the overall transient characteristics are still present.

The same algorithm could be adapted for recognizing a variety of bodily gestures sensed through the floor surface.

3.7 Conclusions

This chapter reviewed approaches to interacting with computationally augmented ground surfaces, for purposes such as those of realizing virtual ground material simulations or foot-based touch surface interfaces. It described a particular approach to realizing such interactions via a distributed, multimodal floor interface, which is capable of capturing foot-floor contact information that is highly useful in the aforementioned settings. It presented interaction methods that are low in cost and complexity, and that can be made accessible to multiple users without requiring body-worn markers or equipment. In addition, this chapter presented examples in which these interaction techniques are used to realize generic virtual control surfaces or immersive geospatial data navigation applications. Further applications to the realization of multisensory simulations of virtual ground surfaces are presented in Chapter 8 of this book. We also reviewed guidelines for the use of such a display, based on a usability evaluation applied within a graphical user interface target selection paradigm. It is thus hoped that this contribution succeeds in demonstrating some potential uses and design considerations for floor-based touch surfaces in virtual reality and human-computer interaction.

Chapter 4
A review of nonvisual signatures of human walking with applications to person tracking in augmented environments

Y. Visell, R. Rajalingham, and J. Cooperstock

Abstract Walking can be considered to generate a rich variety of multimodal information that can be made available to ambient sensing systems. Much of this information can be considered to be non-visual in nature, consisting of acoustic and vibration signatures of walking, foot-ground forces, and bodily movements. This chapter reviews techniques for tracking the movements of pedestrians from such signals. We first review the mechanisms through which the latter are generated, then discuss algorithms that can be used to intelligently filter such signals and to extract information from them, through a review of prior literature on human activity and movement tracking, focusing primarily on non-visual sensing modalities. Finally, we discuss one approach to tracking the movements of pedestrians by means of a force-instrumented floor surface.

4.1 Introduction

Walking is fundamental to our experience of the world, and is relevant to many of the immersive conditions that VR systems aim at simulating. It holds also promise for applications in ubiquitous computing, based on the idea that human locomotion and navigation can function as effective and natural means for interaction with digital information [33, 100].

Most work involving the computational understanding of walking activities has been focused on visual sensing paradigms. However, non-visual channels, consisting of sound, force, and vibrational signatures it produces, are also rich in information. Such signatures are suited to revealing the contact phase of the walking cycle, during which the foot and ground are subject to large mutual forces. Information about this phase is only indirectly reflected in data acquired pertaining to motions of the body, filtered through its inertial mass and structure. As a result, much of the literature on modeling walking dynamics within the vision community has tended to ignore the (relatively nonlinear) contact phase, and has preferred to model loco-motion as a linear (smooth) dynamical process.

Walking can be regarded as a dynamic activity that is highly structured in time and space. This chapter mainly considers the way it is reflected through acoustic, inertial and force signatures, acquired through electronic sensors located in the en-vironment. It discusses recent developments in visual and non-visual (in particular, audio) information processing, and applies techniques from statistical time series modeling. The algorithms that are described here can be used to design novel meth-ods for pedestrian interaction within everyday indoor (e.g., office buildings) and outdoor (e.g., park trails) environments, as reviewed below.

This chapter particularly addresses the modeling of contact dynamics of walking as reflected through its remote signatures. From an application standpoint, the re-sulting systems may be useful for the design of display methods for virtual walking experiences for otherwise neutral surfaces. Such displays interactively synthesize and present, through their actuators, synthetic walking signatures, in the form of au-ditory and vibrotactile stimuli, that would normally associated to real world ground surfaces; see Chapters 2, 3, and 8.

Some examples of end-user scenarios may help to ground the discussion that follows:

- Eldercare patients are able to move freely about a smart home or shared liv-ing facility. The home acquires information about their locations and activities via inexpensive and minimally-intrusive acoustic and seismic sensors in order to provide context-sensitive cognitive assistance, or medical aid, in the case of emergency.
- A visually impaired pedestrian in a noisy urban environment arrives at a cross-walk. Upon stepping on the area of sidewalk near the direction she wishes to cross, she receives an intuitive vibrotactile warning from the ground surface, felt by her foot, indicating that it is not yet safe to do so. A moment later, when the crosswalk indicator changes, she receives a second cue indicating that it is safe to cross.
- Visitors to a historical site or museum are guided toward nearby features of in-terest via virtual trails, supplied by the ground or via footwear. These trails dy-namically form, merging their assumed routes toward potential destinations that are consistent with their position, heading, and any prior information.

4.2 Non-visual signatures of human locomotion

4.2.1 Contact interactions in walking

The recognition or tracking of walking from the the forces, vibrations, and sounds it generates is naturally informed by the characteristics of normal human locomotion and the physical interactions between the foot and ground. This topical area in the movement sciences is extensive, including areas of biomechanics and sensorimotor control in locomotion. A large literature exists on the analysis of gait in terms of body posture and movement over time. Some gross features and terminology surrounding human locomotion are reviewed below.

4.2.2 The human gait cycle

The systematic measurement of human locomotion began near the end of the 19th century, when French scientist E. J. Marey began to study human gait using sequenced multiple camera photography [24]. He developed the first myograph (a device for measuring muscle activity), as well as a foot-switch system the measure the magnitude and timing of planar contact. Analysis of normal and pathological human gait continued to be developed over subsequent decades. Measurement techniques were developed based on motion pictures (film), sensor-instrumented walkways, on-body instrumentation, including electromyographics (EMG) and goniometry, and later using video and motion capture.

Walking is a periodic activity, and a single period is known as the gait cycle. Typical human walking rates are between 75 and 125 steps per minute (about 1.25 to 2 Hz) [237]. It can be divided into two temporal phases – those of stance and swing (Figure 4.1). The former comprises approximately 60 percent of the cycle,

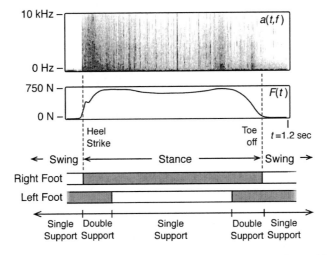

Fig. 4.1: Vibration spectrogram $a(t, f)$ and low-frequency normal foot-ground force $F(t)$ measured at the hard sole of a men's shoe during one footstep of a walker onto rock gravel, together with the corresponding foot contact states within the gait cycle (author's measurements). The dark vertical stripes in the spectrogram correspond to discrete impact or dislocation events that are characteristic of dynamic loading of a complex, granular medium.

during which some part of the walker's weight is carried by that foot, and can be characterized in terms of foot position and contact, decomposed into initial heel strike, followed by foot flat, heel off, knee flexion, and toe off. The subsequent swing phase spans the period from when the toe first clears the ground until ground contact is reestablished. Thus, the gait cycle is characterized by a mixture of postural attributes (e.g., the degree of flexion at the knee) and contact attributes (e.g., toe on or off). Several time scales are involved, including those of the walking tempo or pace, the footstep, encompassing an entire swing-stance cycle, and discrete events such as heel strike and toe slap.

The net force $\mathbf{F}$ exerted by the foot against the ground is a vector with components tangential and normal to the ground surface. One can distinguish between low and high frequency force components, above and below about 50 Hz. The former is referred to as the GRF, and is responsible for the center of mass movement of the individual and the support of his or her weight. Although approximately independent of footwear type, it varies between individuals and walking styles [105]. High-frequency forces are due to material-dependent transient, noise-like, or oscillatory interactions between the shoe and ground surface, and are responsible for airborne acoustic and mechanical vibrations generated during walking [170, 85, 322, 67, 86, 105]. Since these high-frequency signals depend energetically on walking movements that generated them, the gait frequency is readily detected in remote measurements via seismic and acoustic sensing [126, 86, 105, 84]. Signatures of individual footsteps are marked by the two key events: the heel strike and toe slap [170, 85] (Figure 4.1). The time between the two is on the order of 100 ms for stereotypical walking. However, depending on the shoe type, activity, and ground surface shape, a more complex temporal dependence may be observed.

4.2.3 Mechanical interactions and materials

Stepping onto a natural or man-made surface produces rich multimodal information, including mechanical vibrations that are indicative of the actions and types of materials involved. Footsteps in hard-soled shoes onto solid materials are typified by transient impacts due to initial heel strike with the floor surface and subsequent slap of the toe area, while sliding can produce either frictional squeaking (when surfaces are clean) or textured noise. In indoor environments, the operation of common foot operated switches, used for lamps, dental equipment, or other machines, is often accompanied by transient contact events between solid mechanical elements.

The discrete quality of these mechanical signals can be contrasted with the more temporally extended nature of those generated by steps onto natural ground coverings, such as gravel, dry sand, or branches. Here, discrete impacts may not be as apparent, and can be accompanied by both viscoelastic deformation and complex transient, oscillatory, or noise-like vibrations generated through the inelastic displacement of heterogeneous materials [118]. A few of the processes that can be involved include brittle fracture and the production of in-solid acoustic bursts during rapid micro-fracture growth [118, 8, 4], stress fluctuations during shear sliding on granular media [25, 71, 204, 203], and the collapse of air pockets in soil or sand.

Thus, foot-ground interactions are commonly accompanied by mechanical vibrations with energy distributed over a broad range of frequencies. High-frequency vibrations originate with a few different categories of physical interaction, including impacts, fracture, and sliding friction. The physics involved is relatively well understood in restricted settings, such as those involving homogeneous solids, but less so when disordered, heterogeneous materials are involved. Due to the random nature of stress-strain fluctuations in these materials, statistical modeling approaches have seen considerable success in describing them [8, 118]. Chapter 8 discusses the interactive simulation of stepping onto such materials in VR.

4.3 Extracting information from vibrational, GRF, and intertial walking signatures

The processing of information in vibrational signatures of walking has been accomplished using data from a diverse array of sensors, including seismic geophones, MEMS accelerometers, piezoelectric accelerometers or other vibration sensors. As noted above, ow frequency components of such signals (below about 300 Hz) are characteristic of the gait pattern, while higher frequency components are indicative of the footwear and ground material types involved.

Research in the mining of information from such signals has aimed at developing suitable feature extraction methods for these data sources [127, 220]. Wavelet based representations and blind wavelet denoising (a signal processing method for the reduction of noise in data, based on the estimation of noise contributions at different scales) has been applied to seismic footstep data by several prior authors [196, 329]. The identification of promising inference or classification algorithms is another active area of research [40, 287, 286, 283, 239].

Seismic signals in the ground can be detected over long distances given amenable conditions, and this has spurred interest in the analysis of such signals for human surveillance [329, 85, 87]. Further work in this area has addressed human distance estimation from seismic footstep data [219]. Specific methods have been developed for footstep detection via distributed sensor networks in uncontrolled environments [182].

Several research groups have investigated classification of inertial sensing and low frequency foot-ground force profiles (ground reaction forces acquired via force sensing floors), for applications such as biometric authentification. Mostayed, Kim, Mazumder, and Park studied the recognition of walker identity from ground reaction force profiles acquired through a force plate [196], as did Headon and Curwen [115]. Orr and Abowd, and Addlesee [2] performed similar research aimed at the development of a floor for a smart home environment. Several other research groups have carried out work in this area.

Another area of research centers on gait recognition using wearable inertial sensors, or those embedded in handheld computers. Gafurov et al. utilize a sensor pack attached to the lower leg [103] or hip [104]. Ailisto et al. used a sensor pack attached to the waist [7] or the accelerometer of a handheld computer [180]. Both sets of authors experimented with simple, hand-crafted linear discriminant classifiers based on human inspection of the statistics of the data.

One promising application in this area is the early identification of at-risk gait in aging populations from foot-ground force measurements. Holtzreiter and Kohle [125] used artificial neural network to classify normal and pathological gait using force platform recordings. Other researchers have studied the identification of pathological gait patterns using non-contact based sensing. Begg et al. [32] use a support vector machine classifier (SVM) to identify pathological gait using features derived from the minimum foot clearance, which is a local minimum of the vertical distance between the shoe and ground that occurs after toe off in the gait cycle. The required data is typically acquired through motion capture, which has also been used in many other studies on gait classification. Indeed, video based motion capture data likely provides better information for the identification of potential gait pathologies than can be obtained through indirect sensing methods based on non-visual walking signatures.

Chau [55, 56] reviews techniques for analysis and classification of gait data from a movement sciences perspective, with an emphasis on algorithms as opposed to sensing methods. Most examples cited in these surveys utilized motion capture or electromyography for sensing. The volume edited by Palaniswami et al. [222] provides a broad review of machine learning techniques applied to the analysis of human motion from a clinical and movement sciences perspective, emphasizing algorithms and analysis techniques, as distinct from sensing methods and data sources.

4.3.1 Multimodal context and activity recognition

The computational understanding of human activities from multimodal sensors has been widely investigated in ubiquitous and wearable computing research. Some of this research aims at classifying human activities, locations, or other contextual information from several heterogeneous sensor data sources.

4.3.1.1 Multimodal wearable computing systems for activity recognition

Multimodal sensing of walkers movements has also been accomplished with wearable computing devices. One difficulty with evaluating and comparing published results on human activity recognition from wearable sensors is the lack of standard tasks and data sets, and the distinct applications (such as context recognition vs. activity recognition) or metrics (such as retrieval precision and recall vs. classification error rate) that have been addressed, and different data sources that are used.

   Lester and Choudhury address activity recognition and tracking using a wearable multi-sensor unit, including an accelerometer, microphones, light sensors, barometric pressure, humidity, temperature, and compass sensors. They begin with a large number of features (651 in all) calculated from this data, and apply boosting (an ensemble classifier method based on the consistent combination of classifiers [99, 184]) to select a fraction for use in classification by hidden Markov models (HMMs). These HMMs are trained on posterior probabilities computed from simple ("stump") classifiers for each of the selected features. Subramanya et al. adopt a dynamic Bayesian network (DBN) approach to the recognition of activities and spatial context from a wearable sensor system comprising a global positioning system (GPS) unit, accelerometer, microphones, light, temperature, and barometric pressure [282]. The DBN models employed allow their system to model relationships between locations, GPS measurements, state, environment, and other sensor measurements, through time. Dynamic Bayesian networks are generalized probabilistic graphical models structured over time. They can be implemented as Bayesian filters, and special cases include the HMM, the conditional HMM and the Kalman filter. DBNs make it possible to encode a wide range of types of conditional probabilistic dependencies between state variables and observations over time [200].

   Inertial sensors are inexpensive, and increasingly available on commonly used mobile digital devices, such as telephones. They provide a simple way to capture bodily movement. Nonetheless, this data is highly sensitive to the body location at which they are located, which can be addressed through the design of the physical device in relation to the body, and through the extraction of features that can be interpreted despite variation in location. Kunze et al. addressed an unusual problem in this regard: the automatic recognition of the location on the body at which the accelerometer is located [154]. Many researchers have recently studied user activity recognition from accelerometer data [26, 27, 117, 130, 145, 151, 156, 166, 169, 180, 251, 249, 155, 290, 146]. A few recent reviews of the analysis and recognition of human activities via accelerometers exist [181, 10, 148]. Cheng and Hailes investigated the use of accelerometers mounted near the foot for identifying activities performed by their wearer [58]. Huynh and Schiele aimed to extract more informative representations from accelerometer data [130]. They evaluated short-time spectral and average temporal features of body-worn accelerometer data, as well as measures of pairwise correlation between components. Several window lengths were investigated. Evaluation was performed using unsupervised (clustering) and supervised (recognition) methods. Differences were found in feature salience between static and dynamic activities. Heinz and Kunze also assessed features extracted from accelerometer data [117]. Intille et al. have carried out a wide range of research aimed at recognizing activities in domestic environments. Tapia and Intille combined data from wireless accelerometers and heart monitors for real-time recognition of physical activities [289]. Kunze et al. addressed an unusual problem in this regard: the automatic recognition of the location on the body at which the accelerometer is located [154].

   Audio data from microphones is used in mobile and ubiquitous computing to providee structured records of environments and activities in them, being less ob-

trusive than video sensing, and less sensitive to bodily location than video or inertial sensing. Pärkkä et al. studied classification of human movement activities such
as walking, running and cycling from wearable sensors, using 35 body worn sensors [228]. The latter ranged from microphones to physiological sensors, barometric
pressure, temperature, and accelerometers. Short-time temporal and spectral features were extracted from these sources. Additional features were calculated from
stump classifiers designed to identify particular attributes via single sensor modalities (e.g., speech from sound). Features known to be relevant to certain of the measurement modalities (such as physiological sensors) were also calculated. Feature
selection was performed in a heuristic manner, using examination of the statistical
distributions of the features, and domain-specific knowledge. Temporally stationary classifiers, based on manually and automatically constructed decision trees, and
artificial neural networks, were applied to a seven activity classification task. High
accuracy was reported on this task. The authors also report that, contrary to expecta-
tions, physiological signals such as heart rate and respiration did not prove very useful for activity recognition. Korpipää et al. utilized a wearable sensor pack, including microphone sensor, for classifying 19 environmental contexts encountered by its
wearer [150]. Features were derived from the audio data using methods specified in
the MPEG-7 standard, including psychoacoustic features and short time spectrum
representations. Features were subjected to vector quantization, using fuzzy logic
and heuristics. Classification was performed using a naïve Bayes network classifier,
without a model of structure on timescales longer than the analysis frame. Recognition results were reported to be relatively high, using a proprietary dataset. Staeger,
Lukowicz, et al. focus on sound recognition with a low-power, low-cost miniature, wearable acoustic sensor [276, 175]. They extract a set of short-time spectral
and temporal features, perform feature selection using linear discriminant analysis
(LDA) and principle components analysis (PCA), and perform evaluation by classifying the location of the wearer, or the dominant sound source, using k-nearest
neighbors or learning vector quantization (LVQ) classifiers. These classifiers were
selected over more powerful algorithms, due to their suitability for implementation
on low power portable electronic devices. Sound source classification has also been
investigated with the aim of improving hearing aids [50].

4.3.1.2 Walking sound classification

A handful of published studies have reported on the classification of walking sounds.
The subject has arisen in the automatic classification of everyday sounds, walking
sounds typically representing one of several categories to be identified, or in auditory scene recognition, where walking sounds are characteristics of certain scenes.
It has also been studied in the context of biometrics, surveillance and ambient intelligence.

Walking sound classification studies can be distinguished, in part, by the tax-
onomies of types of sound that are employed. Several have focused on differentiating walking sounds from other sound categories. Some address the classification of

walkers according to their unique identities, or attributes such as gender, weight, or footwear. Others attempt to classify walking style or the affective state or intent of the walker. Still others concern themselves with the classification of environmental properties such as ground material or location.

Casey [52] studied the classification of sounds into 19 different categories, ranging from female speech, to telephones, footsteps, and piano sounds, using the MPEG-7 Audio Framework sound classification tools [53]. Features were determined from an ICA decomposition of the log spectral energies of the signals. These were fed to 19 single Gaussian HMMs with full covariance matrices, trained using a Bayesian maximum a posteriori (MAP) procedure. The mean recognition rate on the experiments was more than 92%. Footstep sounds were classified with an accuracy of about 90%, while a few sounds (such as telephones, at around 65%) were mis-classified much more often.

Mitrovic [189] studied combinations of features and classifiers for everyday sound retrieval. Five environmental sound categories were included , labeled "cars", "crowds", "footsteps", "signals", and "thunder". Data was taken from public internet sources. This study evaluated psychoacoutic features, time-domain measures, and spectral features, such as linear predictive coding (LPC) coefficients and Mel Frequency Cepstral Coefficients (MFCCs). Classification was performed with three classifiers: the k-nearest neighbor algorithm, learning vector quantization, and the support vector machine. Retrieval recall and precision were highest with MFCC features in combination with a support vector machine classifier. The author also applied feature selection (based on a posteriori retrieval performance) to determine promising combinations of features.

Peltonen et al. [235] and subsequently Eronen et al. [90] studied the classification of real-world acoustic environments from audio recordings, unlike many other studies on everyday sound classification that use composed studio recordings of different soundscapes. Several of the environments in their study (e.g., indoor hallways) included walking sounds, although none was uniquely characterized by the presence of walking. 27 contexts and six high-level categories were represented. The authors studied recognition performance as a function of features, feature selection methods, HMM configuration and training method. They achieved their best results using PCA or ICA transformation of MFCC features, and found that performance remained near its highest level (around 58% for context recognition) even when a low (four-) dimensional feature vector was used. Their results were compared with human performance on the same tasks, and they found that humans performed only moderately better (at 69% on the context recognition task). Several other authors have studied everyday soundscape classification [18, 17].

A number researchers have investigated features and classifiers for biometric authentification or walker identification from footstep sounds. Many of the results presented in the literature to date are limited by the use of small or studio-recorded datasets. Itai and Yasukawa [134] focused on psychoacoustic parameters such as loudness, roughness, and fluctuation strength, as well as MFCC features, using a vector quantization based classifier. They reported 88% recognition rates on an identification experiment with 5 different walkers. The same authors later investigated

classification based on dynamic time warping [136] and wavelet based features
[135]. Shoji et al. [270] and Tanaka and Inoue [288] also carried out classification
experiments using spectral features without time information.

Annies et al. [11] studied the automatic classification of walking sounds according to the footwear or the ground material type (snow, sand, metal, etc.), using
recordings from a commercial sound design database. They compared classification
performance using two feature types and two classifiers. Both features were derived
from an auditory signal processing model due to Patterson and Holdsworth [229].
One was oriented toward capturing fine temporal information in the sounds, and
the other toward short-time spectral information. An HMM classifier was trained
using the feature sequences for the different recordings, while the SVM classifier
was trained using aggregated features without sequential information. In nearly all
cases, the SVM approach outperformed the HMM.

Bland [40] explored simple methods for detecting footstep events via seismic
sensors and microphones. An autoregressive process (i.e., an all-pole IIR filter) was
adapted on each window of an overlapping series of windowed footstep recordings,
and rough examination of the resulting coefficient paths was used to guide a qualitative analysis of differences between footstep types.

Abu-El-Quran et al. [1] developed a microphone array surveillance system. Their
system includes a module for classifying non-speech audio using MFCC features
and a time-delay neural network, together with a classifier for distinguishing speech
and non-speech audio. Evaluation was performed using a database of recordings
acquired under controlled (studio) conditions. They report average error rates of
around 5% on a five class classification task including speech, door sounds, fan
noise, glass breaking, and wind. Footstep sounds were evaluated only in the context of speech-nonspeech audio classification, where they were never confused with
speech.

Itai et al. investigated psychoacoustically-motivated features for the classification
of walker identity, in a study in which walkers performed on a wood floor inside a
house [134]. In the context investigated, the authors reported mixed success in clustering walker identities within a multidimensional space whose dimensions were
loudness, sharpness, fluctuation strength, and roughness.

4.3.1.3 Ambient sensing of human activities using acoustic information

Zou and Bhanu studied the tracking of walkers' location using walking-generated
acoustic signals captured from a pair of microphones, as well as video input [35].
Zou and Bhanu found that a (stationary) Bayesian network outperformed a time-
delay neural network (which is able to model temporal correlations explicitly) on a
walker localization task.

Radhakrishnan and Divakaran applied generative background models normally
used for visual background modeling, to the modeling of background audio processes. Their approach, based on Gaussian mixture models and short-time cepstral
features, was applied to an audio surveillance problem aimed at novelty detection

[246]. Lukowicz, Ward et al. utilized a set of body worn microphones and accelerometers for recognizing activities performed by their wearer over time, in a wood workshop context [176, 320, 321]. Classification decisions were formed using the output of separate models for audio and accelerometer data. Microphone pairs were used to distinguish others' activities from those performed by the wearer. Minnen, Starner, Lukowicz, et al. focused on the recognition of human actions, at the level of individual gestures, using similar data [188]. Chen et al. developed and evaluated a system for monitoring bathroom activities from acoustic information. Their method employed Mel-frequency cepstral coefficient audio features with an HMM classifier, an approach that is typically used for automatic speech recognition [57].

4.3.1.4 Acoustic source separation and localization

Acoustic source localization a long history in SONAR and ballistics imaging, and related research has been carried out in RADAR localization. The localization of humans via such information has been most comprehensively addressed in the literature on speaker tracking from microphone arrays [330, 167, 241, 258], either in a unimodal or multimodal setting (the latter typically also utilizing video sensing).

Interest in distributed sensor networks and ubiquitous computing has led several researchers to study the localization and identification of human activities from acoustic information captured through sensors embedded in everyday artifacts or environments. Dalton explored algorithms for passive acoustic source localization using indoor sensor networks [70]. Westner and Bove examined the separation of acoustic events using microphone arrays in rooms, using blind source separation techniques [326, 325]. Acoustic source localization also holds potential for improving computational auditory scene analysis systems [88]. Weinstein et al. report on the development of a large scale (1020-node) microphone array and acoustic localization system (beamformer) for intelligent environment applications [324]. Other researchers have approached acoustic source localization using wearable computing (and microphone array) devices, for audio annotation and retrieval [30, 64], or for situation analysis by auditory scene recognition [235].

4.3.1.5 GPS-based activity tracking

Contact-based sensing of walking activities may be applicable to settings in which other sources of data, such as GPS, have traditionally been used for human activity tracking [173, 171, 14]. Due to the coarser spatial resolution of GPS data, activities recognized are typically long in time scale. Ashbrook and Starner focus on movement between locations, without considering the activity context or route choices involved [15]. Bennewitz et al. demonstrate learning of motion paths between places. Liao et al. address the problem of annotating a user's daily activities [173], and that of jointly modeling place and activity from such data [172].

4.3.1.6 Interface design with acoustic sensing

Murray-Smith and Williamson developed an input method based on real time classification of acoustic information generated through contact events between a user's fingers and the plastic case of a handheld information appliance, using data from inexpensive acoustic sensors, features derived from the magnitude short-time Fourier transform (STFT), and a neural network classification algorithm [201]. A few research projects, including the European Commission sponsored TAI-CHI project, have investigated the extraction of position information from human contact with flat surfaces, using array based signal processing techniques applied to in-solid acoustic signals acquired from contact microphones [68].

4.4 Walking signatures: Dynamic information processing

As reflected in the scenarios and choice of examples presented above, one theme of this chapter is that of modeling pedestrian positions and activities as they evolve over time. The attributes that might be used to describe such pedestrian states are not only time-varying, but express intrinsic temporal structure and dynamics, due from the physical constraints that determine the biomechanics of locomotion, the (discontinuous) temporal evolution of foot-ground forces through the gait cycle, and the physics of contact interactions between foot and ground. Thus, one should not assume that a static classification task is to be solved, but rather should benefit from the structured nature of these interactions. Models suited to systems that evolve through time have been progressively developed in the machine learning community during recent decades.

Two problems relevant to this research concern tracking pedestrian states (activities or positions) over time, and constructing prior models of their activities or trajectories from observations. Here, observations are assumed to consist of contact-generated walking signatures. The term *pose* is used in the literature on computational perception to refer to the complete configuration of a system of interest, including any unobserved aspects of its state. A typical problem to be solved is that of determining pose from partial observations. This may mean reconstructing a complete description of the position or activity of one or more individuals. Such problems arise in many location-aware applications of pervasive and ubiquitous human computer interaction. The problem is typically addressed by combining information from continuously updated data, acquired from sensors, to make inferences about the expected position or activity of the individual. Such estimates are improved if a good prior model of the behavior or dynamics of the individual is available to guide estimation. This section introduces algorithms and models that are applicable to such a setting.

4.4.1 Probabilistic inference with dynamical models

State space models provide a compact framework for describing the evolution of a dynamical system through time, and they can be elegantly integrated with probabilistic machine learning methods for pattern classification, as shown below.

4.4.1.1 Discrete time dynamical system formulation

The goal is to model time series of walking activities. The time series is represented by a sequence of hidden states x_t (p-dimensional vectors), inputs u_t (r-dimensional vectors), and outputs y_t (k-dimensional vectors). The system evolves in discrete time (the subscript t is the time index) according to a stationary Markov dynamical system:

$$x_{t+1} = f_\theta(x_t, u_t; w_t) \tag{4.1}$$

$$y_t = g_\theta(x_t, u_t; v_t) \tag{4.2}$$

The functions f_θ and g_θ are vector valued, possibly nonlinear functions of their arguments. Often they are parametrized by a set of values θ.[1] w_t and v_t are noise processes. In the simplest nontrivial case, these are taken to be zero mean additive Gaussian processes (Figure 4.2). The system is assumed to be stationary, in the sense

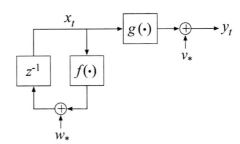

Fig. 4.2: Discrete time non-linear dynamical system model, assuming additive noise. The control input is suppressed.

that neither the noise statistics or the functions f_θ and g_θ depend on time. In case these quantities depend on time, it is preferable to represent this time dependence in the system state. The following definitions are needed:

- x_t: The state vector at time t
- u_t: The (known) control vector, if present, used to drive the state from x_{t-1} to x_t
- y_t: The observation of the state x_t taken at time t
- The state history to time t, $x_{1:t} = (x_1, x_2, \ldots, x_t)$
- The observation history to time t, $y_{1:t} = (y_1, y_2, \ldots, y_t)$
- The similarly defined history of control inputs (if present), $u_{1:t}$

---

[1] Interesting nonparametric nonlinear state space models also exist.

y_t are the observations of what is taking place; here, they will be observations of pedestrian positions, activities, or other walking events, as captured through contact-based sensors. x_t are hidden variables responsible for the observed variation; in this chapter, these will represent pedestrian positions or activity labels. Depending on the inference problem, the inputs u_t may or may not be present. Since it is assumed that walkers act autonomously, u_t will not typically be relevant.

Equation (4.2) is a parametrized control system model of the dynamics of interest. Separating the system state dynamics and observation descriptions provide a conceptually clearer, and in many cases simpler, description. The system may be viewed as a probabilistic latent variable model, in which the states x are thought of as hidden variables, furnishing an informative lower dimensional explanation of complex variations observed in the outputs y. Equation (4.2) can be thought of as specifying a dynamical prior probability distribution over the latent space. A unified view of such models was presented by Roweis and Ghahramani [259]. The form and parameter values of the dynamical function f_θ and the observation function g_θ, together with the noise distributions, determine the qualities of the observed time series y_t. The system dynamics depend both on the dynamical function's specification and on the state description. The former may be inferred from data, or based on a priori considerations, such as physical models.

4.4.1.2 State and model estimation from observations

Assuming an observed input-output time series (u_t, y_t) to be given, and assuming the parametrized form f_θ, g_θ of the system to be appropriate, one would like to determine values of the system parameters θ that best explain the observations, providing a compact representation of the observations. The problem of determining the values of θ in this way is called *model estimation*. (*System identification* is a name commonly used in the engineering literature.)

If a time series model has been estimated from a sequence of observations up to the present time (which can be compactly denoted as $y_{1:t} = \{y_1, y_2, \ldots, y_t\}$), one can ask what unobserved state sequence $x_{1:t}$ is implied. More generally, one may wish to know how well a given state sequence fits with the observations, best addressed within a probabilistic setting (Section 4.4.1.3). This is the *state estimation* problem (also referred to as *tracking, filtering* or *smoothing* in the engineering literature). Tracking and filtering attempt to infer the hidden state at time t from observations up to time t, while smoothing takes into account all available observations, up to and beyond time t. As an example, y_t might be noisy measurements of a walker's position, from which one wishes to estimate the true spatial position x_t.

Various techniques have been developed over the last half century for solving state and model estimation. Many of them can be understood from the viewpoint of Bayesian filtering, a probabilistic framework for updating such predictions continuously as new information is acquired from one or more sources.

4.4.1.3 Probabilistic formulation

Realistic dynamical system models have a stochastic quality, because the systems of interest have observations that are noisy functions of the inputs, and possess noisy dynamics. In such a model, process noise w_t and observation noise v_t are meant to reflect uncertainty in observed and hidden state space variables. An additional advantage of noise is that a wide variety of system outputs can be captured by a simple model (avoiding, for example, the need to explicitly modeling complex microscopic disturbances).

The system state in the presence of noise can be represented through the use of probabilities, which reflect the uncertainties involved, and provide a calculus for updating state estimates and uncertainties as new information is received [96, 296]. A common approximation is to assume that v_t and w_t represent additive, white (temporally uncorrelated), zero-mean Gaussian distributed noise sources, i.e., that the noise sources are distributed as $p(v_t) = N(0,Q)$ and $p(w_t) = N(0,R)$, where $N(\mu,\Sigma)$ is a multivariate normal probability density with the indicated mean and covariance matrix, and Q and R are the noise covariances of v_t and w_t. Under this assumption, and suppressing the control inputs, the system equations (4.2) can be written in terms of the conditional densities for the state and output:

$$p(x_t|x_{t-1}) = \exp\left\{-\frac{1}{2}[x_t - f_\theta(x_{t-1})]^T Q^{-1}[y_t - f_\theta(x_{t-1})]\right\} \det(2\pi Q)^{-1/2} \quad (4.3)$$

$$p(y_t|x_t) = \exp\left\{-\frac{1}{2}[y_t - g_\theta(x_t)]^T R^{-1}[y_t - g_\theta(x_t)]\right\} \det(2\pi R)^{-1/2} \quad (4.4)$$

T in the exponent denotes the transpose, and $\det(A)$ is the determinant of A. This set of equations has the interpretation of a dynamic probabilistic latent variable model, with potentially nonlinear dynamics (Equation 4.3) and latent space embedding (Equation 4.4). If f or g are assumed to be linear, the respective densities are Gaussian under the noise assumptions given above. If both are linear, the result is a dynamic, probabilistic linear latent variable model – a type that has been well-studied.

The Markov assumption means that the joint probability can be factored

$$p(x_{1:t}, y_{1:t}) = p(x_1) \prod_{t=1}^{T} p(x_t|x_{t-1}) \prod_{t=1}^{T} p(y_t|x_t). \quad (4.5)$$

If the initial state density is assumed to be a Gaussian $p(x_1) = N(\pi_1, V_1)$ with mean π_1 and covariance V_1, then the joint log probability can be written as a sum of terms having the following form [108]

$$
\begin{aligned}
\log\{p(x_{1:t}, y_{1:t})\} = & -\sum_{t=1}^{T} \left(\frac{1}{2}[y_t - g_\theta(x_t)]^T R^{-1}[y_t - g_\theta(x_t)] \right) - \frac{T}{2}\log|R| \\
& -\sum_{t=2}^{T} \left(\frac{1}{2}[x_t - f_\theta(x_{t-1})]^T Q^{-1}[y_t - f_\theta(x_{t-1})] \right) - \frac{T-1}{2}\log|Q| \\
& -\frac{1}{2}[x_1 - \pi_1]^T V_1^{-1}[x_1 - \pi_1] - \frac{1}{2}\log|V_1| - \frac{T(p+k)}{2}\log 2\pi \quad (4.6)
\end{aligned}
$$

This (nonlinear) expression is general under the Gaussian noise assumptions above. If the system is assumed to be linear, the log probability is a sum of quadratic terms. Often, neither assumption holds.

4.4.1.4 Forecasting and belief

A basic question concerns how likely any given future observation is, given past observations and (known) control inputs. It is embodied by the forecasting distribution $p(y_{t+h}|y_{1:t}, u_{1:t+h})$. h represents a prediction time horizon. Another quantity of interest is the total probability of an observation sequence given the model, $p(y_{1:T}|\theta)$, (written here to explicitly show the dependence on the model parameters θ).

The state space model assumes the existence of the hidden state x_t whose evolution explains the observations. Computing interesting quantities such as $p(y_{1:T}|\theta)$ is aided if the value of the hidden state of the system can be determined under different conditions. This goal can be embodied by the aim of computing a *belief distribution*, which is given by the posterior density over x_t conditioned on the prior observations $y_{1:t}$ and control inputs $u_{1:t}$

$$
\text{Bel}(x_t) = p(x_t|y_{1:t}, u_{1:t}) \quad (4.7)
$$

The belief distribution embodies the knowledge of the system at the present time t given all past observations and control inputs.

As Murphy notes [200], a rationale for belief representations as the objective for estimation using stochastic dynamical systems was provided in the 1960s, when Aström [16] proved that $\text{Bel}(x_t)$ is a sufficient statistic for prediction and control, provided the hidden state is rich enough. The implication is that nothing that is not already reflected in the belief state at time t is gained by retaining a longer record of past observations.

Since this chapter is not primarily concerned with control, control inputs u_t to the system will be suppressed during the rest of the discussion.

4.4.1.5 Statistical decision making

One aim of this research is to provide models that are both interpretive and that are capable of enabling statistical decision tasks. Interpretive models allow to make inferences as to the continuous states associated to a pedestrian, while decision tasks permit inferences of a discrete nature to be made, such as classifications of identity, activity and gait type. One advantage of the use of probabilities is that they provide a normalized measure of fitness, namely the total likelihood of the observation sequence given the model, $p(y_{1:T}|\theta)$. Although explicitly discriminative approaches, which would provide additional comparisons of relative fitness between models, are not discussed here, such approaches can be developed within the same framework.

4.4.1.6 Sequential state estimation by belief propagation

The belief distribution is influenced by evidence represented by the observation history $y_{1:t}$. The Markov assumption allows to iteratively compute $\mathrm{Bel}(x_t)$ (given an amenable system and belief representation) as the number of sensor observations that have been received grow over time. Inference is accomplished via Bayes' rule[2], cast in a form suitable for recursive state estimation:

$$p(x_t|y_{1:t}) = \frac{p(y_t|x_t,y_{1:t-1})p(x_t|y_{1:t-1})}{p(y_t|y_{1:t-1})} \tag{4.8}$$

using the Markov assumption on y_t and the fact that the denominator is a state-independent value, one can write this as

$$p(x_t|y_{1:t}) = C p(y_t|x_t)p(x_t|y_{1:t-1}) \tag{4.9}$$

(The state-independent value C ensures that the posterior distribution integrates to one over the state space.) The last factor can be expanded through the intermediate state x_{t-1} to give

$$p(x_t|y_{1:t}) = C p(y_t|x_t) \int p(x_t|x_{t-1})p(x_{t-1}|y_{1:t-1})dx_{t-1} \tag{4.10}$$

Using this formula, belief is updated at each time step, in two stages. Upon a state transition to x_t, it the **prediction** formula is applied, combining the posterior density $\mathrm{Bel}(x_{t-1}) = p(x_{t-1}|y_{1:t-1})$ at the prior timestep with the system dynamics distribution $p(x_t|x_{t-1})$:

$$\mathrm{Bel}^-(x_t) \leftarrow \int p(x_t|x_{t-1})\mathrm{Bel}(x_{t-1})dx_{t-1} \tag{4.11}$$

The superscript $(-)$ denotes that the belief is a prediction for the future which does not account for the current observation. When an observation y_t is made, the estimate is fused with the new information, via the **correction** formula:

$$\mathrm{Bel}(x_t) \leftarrow C p(y_t|x_t)\mathrm{Bel}^-(x_t) \tag{4.12}$$

---

[2] Bayes' rule: $p(x|y) = p(y|x)p(x)/p(y)$

4.4.2 State spaces, belief representations and state inference

The foregoing formulation is abstract, and how it is implemented depends on the system dynamics, observation process, statistics, and representations of $Bel(x_t)$ and $p(y_t|x_t)$. For certain models, sequential state estimation is tractable, through the calculations given above.

Belief representations can be categorized according to assumptions on the state space, belief state, and other factors [96, 186]. They include:

A. Continuous belief distributions

Gaussian distributions with linear dynamics: The belief takes the form of a normal distribution

$$Bel(x_t) = N(x_t; \mu_t, \Sigma_t) \tag{4.13}$$

Optimal sequential state estimation can be solved exactly, and is accomplished by the Kalman filter update equations [142]. The process is tractable because Gaussian distributions are closed under joint conditioning and marginalization (that is, multiplication and integration).

Gaussian distributions with linearized dynamics: A nonlinear system can often be treated as locally linear. Linearization is accomplished by first order Taylor expansion around the current state. The Kalman filter is applied to this locally linear model. Additional errors arising from linearization are not modeled in the noise variance. The result is called the extended Kalman filter (EKF) [316].

Gaussian distributions with statistical linearization: This model, known as the unscented Kalman filter (UKF), represents a Gaussian state distribution through a set of $2L+1$ weighted samples $x_t^i, i = 1, 2, \ldots, L$ (L is the state space dimension), called *sigma points*, that encode its sufficient statistics. These samples are propagated through the true nonlinearity, and are used to reconstruct an updated estimate of the state distribution via mean and covariance data computed from the propagated samples. The same procedure is applied to the observation model. The UKF can be implemented as efficiently as the EKF, is simpler to implement (requiring no Jacobians to be calculated), and has proved to be more accurate [317].

Mixture of Gaussians: The belief representation consists of a weighted sum of M hypotheses, each of them a normalized Gaussian distribution:

$$Bel(x_t) = \sum_{i=1}^{M} w_i N(x_t; \mu_t^i, \Sigma_t^i) \tag{4.14}$$

Methods in this family can be distinguished according to how they propagate the density forward in time, and how the hypotheses are weighted or selected at successive time steps. Many proposed algorithms are based on heuristics. One configuration consists of M linear Kalman filters, with a prescription for reweighting and/or reallocating filters after temporal updates.

B. Discrete belief distributions:

Monte-Carlo (particle) filters: The state distribution is modeled by a set of discrete samples. They are capable of capturing linear or nonlinear dynamics, and of operating within a continuous or discrete state-space. While similar to weighted mixture of Gaussian models, in Monte Carlo approaches maintaining an optimal set of hypotheses can be done in a theoretically justified way, using importance resampling [13].

Sampled state spaces: Dynamics in such models can be constructed to take place on a regular grid. Since the domains are all discrete, probabilities may be represented by histograms, which makes the relevant computations tractable.

Discrete and nonmetric state spaces: The best-known example is the hidden Markov model (HMM), which assumes that x_t is a discrete random variable. The transition function is no longer continuous, the observation function at any state is any continuous density, and state inference is efficiently performed using the Viterbi algorithm [245].

C. Hybrid continuous-discrete models:

Switching linear dynamical systems can represent systems that are linear in distinct, discretely-parametrized regimes. A discrete state variable (analogous to the discrete state of an HMM) may be used to parametrize the current regime, while a continuous state variable is used to describe dynamics in that regime. Discrete state transitions are typically treated as Markov chains. Examples of systems of this type include models of animal behavior, where several distinct sub-behaviors are observed (e.g., sequences of patterns of movements in populations of bees), or human juggling [212]. For Gaussian-distributed belief representations, learning can be performed through iterative approximation, using the EM algorithm [109]. This family of models has been applied in diverse disciplines, ranging from control, to geosciences, economics, and aerospace engineering, and consequently a broad range of approximate inference methods have been developed.

4.4.3 Locomotion as a piecewise-continuous process

Systems of the last type described above are relevant to the subject matter of this chapter, because of the discontinuous (or, better stated, piecewise-continuous) nature of foot-ground interactions during locomotion. In related research by Bissacco et al. kinematic data acquired during locomotion was viewed as having been generated by a hybrid continuous-discrete dynamical system [39, 38]. In their models, kinematic gait patterns are captured through a switching linear dynamical system consisting of a continuous system (an autoregression process, as described in Section 4.4.4) designed to capture the motion of the leg during the free swing phase, and a discrete transition, which provides a simplified representation of the nonlinear phase during which foot and ground are in contact. These distinctive properties of a gait pattern are regarded as arising from different parameter settings of the continuous model, and from a different discrete transition process. The model estimation and tracking algorithms these authors developed are reviewed in Section 4.4.4, below.

By comparison, the present chapter focuses primary attention on the contact phase. Time evolution within this phase is modeled through observations of the acoustic and vibrational signatures that it produces. The non-contact phases can be viewed as discontinuities during which no observations are available, as footstep signatures are not being produced.

4.4.4 Model estimation from data

The discussion above focused on the problem of state inference, i.e., the consistent calculation of belief distributions about the hidden system state as a function of observations. From a learning standpoint, the complementary problem is that of determining the system model f, g that best fits the evidence. For a parametric system model, this means determining the value of the system parameters θ. The approach taken to learning depends substantially on the model form that is assumed. The two most common families of model estimation algorithm rely on gradient descent optimization or the expectation-maximization (EM) algorithm. Other approaches to learning system models from data, which do not directly fit within these two categories, have also been developed in recent years.

4.4.4.1 System identification

Input-output time series models, such as autoregression processes, and certain black-box models, such as time-delay neural networks, have predominantly relied on gradient based methods for learning model parameters from data.In this form of parameter estimation, model parameters are directly adapted via iterative gradient descent so as to best explain the observed time series Y. An error function is defined between the predictions of the model and the observations – for example, between the average one-step predicted observations and the subsequent true observations, given the model parameters θ. A gradient descent algorithm is used to find the values θ^* of theta that yield a local minimum of this error function.

More modern versions of system identification, that seek to learn parameters of state space system models from input and output observations in a non-probabilistic

setting also exist, but a comprehensive review of other methods from system identification is not attempted here.

4.4.4.2 Expectation maximization

The expectation maximization (EM) algorithm is applicable to the estimation of
a wide class of probability models. It provides a method for determining the pa-
rameters θ that maximize the log likelihood of a set of observed data $\log p(Y|\theta)$,
without knowledge of the values of the hidden state variables X [75]. It assumes a
joint distribution $p(Y,X|\theta)$ over the observed and unobserved variables to be given,
dependent on θ. The algorithm proceeds as follows [37].

1. Choose an initial value of the parameters, θ_{old} .
2. **E-Step:** Evaluate the posterior distribution of the observations, $p(Y|X, \theta_{\text{old}})$.
3. **M-Step:** Determine θ_{new} to maximize the complete data log likelihood $Q(\theta, \theta_{\text{old}})$

$$\theta_{\text{new}} = \arg\max_{\theta} Q(\theta, \theta_{\text{old}}) \tag{4.15}$$

where
$$Q(\theta, \theta_{\text{old}}) = \sum_{Y} p(Y|X, \theta_{\text{old}}) \log p(Y,X|\theta). \tag{4.16}$$

4. Check for convergence of Q or of θ. If convergence is not met, let $\theta_{\text{old}} \leftarrow \theta_{\text{new}}$ and
   return to step 2.

Each cycle of the algorithm is guaranteed to increase the likelihood of the observed
data. When the EM algorithm is applied to time series model estimation [271], the E
step is solved using state estimation to infer the distribution over the hidden state sequence. The M step uses the result of the state estimation step to optimize the values
of the parameters. Applied to the HMM, this results in the widely used Baum-Welch
training algorithm [31]. The algorithm can also be used to learn the parameters of
a linear dynamical system (Kalman filter) [109]. In this case, state estimation in the
M step is performed using Kalman filtering or smoothing.

Several authors have applied the EM algorithm to the estimation of nonlinear
state-space models. Roweis and Ghahramani [260] combined the EKF and EM algorithms to estimate the parameters of a model consisting of a linear-Gaussian system combined with a parametric nonlinearity given by a radial basis function (RBF)
network, i.e., a state-space function

$$z = \sum_{i=1}^{I} h_i \rho_i(x) + Ax + b, \quad \rho_i(x) = \det(2\pi S_i)^{-\frac{1}{2}} \exp\left\{ -\frac{1}{2}(x - c_i)^t S_i^{-1}(x - c_i) \right\}. \tag{4.17}$$

The authors parametrize this nonlinearity as a superposition of Gaussian basis functions, characterized by a set of weights h_i, centers c_i in state space, and covariances
S_i. The same form of nonlinearity is used to represent the system evolution function
f and/or the observation function g, with appropriate replacements of x and z for x_t,
x_{t-1}, and y_t.

4.4.4.3 Bayesian multimodal fusion

When data from multiple sensor channels is available, it is desirable to combine it in ways that allow to improve state estimates. This can be accommodated within the Bayesian filtering viewpoint [83]. Assume the observation vector y to consist of data derived from k distinct sources $y_t = (y_t^{(1)}, y_t^{(2)}, \ldots, y_t^{(k)})$. Provided these measurements are conditionally independent [83], i.e.

$$p(y_t^{(1)}, y_t^{(2)}, \ldots, y_t^{(k)} | x_t) = p(y_t^{(1)} | x_t) p(y_t^{(2)} | x_t) \cdots p(y_t^{(k)} | x_t) \qquad (4.18)$$

then the y's form what is known as an independent likelihood pool,

$$p(x|y) = C p(x) \prod_{i=1}^{k} p(y^{(k)} | x) \qquad (4.19)$$

and the state and model estimation discussions presented above go through without modification. Thus, one major advantage of the Bayesian filtering viewpoint is that multisensor data fusion can, given amenable measurements, be handled within a unitary formalism.

4.4.4.4 Other approaches to model estimation for human motion

Ijspeert et al. [132, 131] developed a state space dynamical system model for human motion, called Dynamic Movement Primitives (DMP). These models consist of one of two canonical linear systems, reproducing either discrete movements (with bell-shaped velocity profiles similar to typical trajectories seen in human point-to-point movements) or periodic, rhythmic movements, based on limit cycle attractors. These basic linear models are augmented with a locally-weighted RBF nonlinearity that can be learned from data to fit a demonstrated kinematic trajectory. From a biological movement control perspective, the DMP is thought of as an elementary predictive (forward) model for quickly executing a previously learned movement in a way that is adapted to the requirements of context (for example, with variation in execution speed, scale, or endpoints of the trajectory). The approach estimates model parameters by integrating the hidden state using the linear part of the model matched to the gross features of the trajectory (e.g., start and end configuration), and applies a form of subspace regression to fit the RBF nonlinearity to the observed kinematic trajectory.

Visell et al. extended the approach of Ijspeert et al. to a Bayesian filter framework by augmenting the dynamical model with a Monte-Carlo sampled belief distribution. The latter is used to perform state inference for new observation sequences [306], and, within a multiple motion models framework, to maintain a continuously updated belief state over the joint set of motion possibilities, which can be used as a filter for interpreting the observed motion pattern in real-time.

Bissacco [39] developed a closed-form, but sub-optimal, model learning algorithm for switching linear dynamical systems, based on subspace system identification ideas. In this model, each linear system is conceived to correspond to the non-contact phase of a gait pattern, while the switch is meant to correspond to the ground-contact phase. This algorithm estimates both model-switching event distributions, and parameters of the models themselves, from data. The model was applied to human gait dynamics, as represented in joint angles acquired via motion capture data. It was found to compare qualitatively and quantitatively favorably to iterative learning methods. However, decision-making tasks such as classification were not implemented.

Bissacco and Soatto [38] developed hybrid continuous-discrete autoregressive (AR) models for human motion. They apply a Bayesian approach to learning posterior distributions $p(\theta|y_{1:t})$ over the parameters of continuous (non-switching) AR models, and develop a model distance using these posterior distributions, based on the Wasserstein-Mallows ("Earth-mover's") distance. Extending this model to the switching case (hybrid autoregressive models) involved applying the distance measure to multimodal distributions. They use their model to learn time series derived from video sensing of human locomotion patterns, and demonstrate the superior classification performance of the resulting hybrid models over simpler linear models. In subsequent work they have developed kernel-based distance measures for the space of autoregressive processes, to support statistical decision making among estimated models (e.g., classification of learned time series). As a question of style, these methods are more closely linked to stochastic control, whereas the foregoing discussion has emphasized a probabilistic viewpoint.

Others have developed approaches to motion dynamics modeling based on Bayesian models with Gaussian Process priors that are capable of jointly performing dimensionality reduction and state space model regression. Wang et al. [319, 318] developed a machine learning model of human movement known as the Gaussian process dynamical model (GPDM). As in the model of Roweis and Ghahramani, the GPDM represents a movement pattern as a probabilistic dynamical model consisting of a linear Gaussian system with an RBF nonlinearity. However, whereas Roweis and Ghahramani learn a parametric model from data, using the EM algorithm, Wang et al. adopt a fully Bayesian approach, and marginalize over (integrate out) the model parameters, with a Gaussian process prior. The marginalization is performed explicitly in closed form, leading to a nonparametric model which can be learned from data. They trained the GPDM on high dimensional motion capture sequences of human walking and evaluated the results of learning using different training algorithms. Related approaches have been developed based on regression with reproducing kernel Hilbert space models [247].[3]

---

[3] These approaches are promising but would derail the present proposal too much to discuss here.

4.5 Application to pedestrian tracking via in-floor force sensing

This section presents a probabilistic approach to the tracking and estimation of the lower body posture of users moving on foot over an instrumented floor surface of the type described in Chapter 3. We use force measurements from the floor to track body posture in 3D space using Bayesian filters with a switching state-space model. Potential applications of this work to person tracking and human-computer interaction are briefly discussed.

4.5.1 Background on 3D person tracking

3D human posture tracking is a classic challenge in computer vision and pattern recognition. Computer vision techniques have been the most widely used for this purpose [190, 191], although several researchers have investigated human tracking via in-floor sensor arrays [199, 216, 3]. Common challenges of person tracking in these domains include the loss of 3D pose information through observation via the sensor array, and the underlying complex dynamics of human motion. Missing information plays an important role in both settings. Losses due to occlusion in video-based tracking are somewhat analogous to the loss of observations while feet are out of contact with the ground during tracking via a floor-based array. Overcoming such missing information has, in part, motivated the approach presented here.

Although motion capture and video can provide high-resolution 3D position information for human tracking, they are not always available and are prone to visual occlusion. Furthermore, state of the art methods for inferring human contact interactions from video provide inaccurate information about interaction forces between body and ground [48]. Such forces are highly characteristic of the individuals and activities generating them (as evidenced in the references given below).

Bayesian filtering provides a unifying view of diverse probabilistic tracking methods. It has been extensively applied to problems in object, person, or context tracking [95, 295]. The effectiveness of such methods stems from their ability to integrate information acquired over time in ways that respect the structure and dynamics of the object or individual being tracked. Nonparametric Bayesian filtering techniques, like the particle filter based model used here, make it possible to perform tracking without making unnecessary assumptions about the form taken by those distributions or the dynamics governing them, by maintaining many hypotheses in parallel.

The system presented here consists of a posture tracking system based on a distributed, sparse in-floor sensing array. Prior literature on the analysis of foot-floor contact forces has addressed applications to immersive interactive media [227], pedestrian identification [216, 285], gait and dance analysis [238], among others [331, 266, 194, 185]. Here, similar data is used for 3D kinematic tracking of users' lower bodies, a task that has received less attention in prior research on floor-sensing arrays. Murakita et al. utilized a Markov Chain Monte Carlo method to track pedestrians via an array of binary pressure sensors [199], albeit with much larger errors (typically 0.6 m) than what we achieve here (Section 4.5.5). Yin and Pai tracked whole-body movements via a high-resolution (and costly) floor sensing array, based on the similarity of force patterns to those recorded in a database of movements [331]. However, their system was limited to tracking relative to a predefined set of static poses.

Our intended applications involve both person tracking and interaction with floor-based touch displays via the feet [314]. In both cases, we are interested in tracking the 3D kinematics of the lower body of moving persons, with an emphasis on the locations of their feet. Although existing force sensing arrays can provide accurate information about foot-floor contact forces, the achievable resolution is often limited by cost constraints. In addition, mappings from patterns of foot-floor forces to body posture are complex and one-to-many. Effective use of prior knowledge about body structure, movement, and walking mechanics is required in order to track posture accurately.

4.5.2 System configuration

The sensing floor used in the experiments is described more fully in Chapter 3 (Figure 4.3). It consists of a 6×6 array of rigid tiles, 30 cm on each side. Each tile is

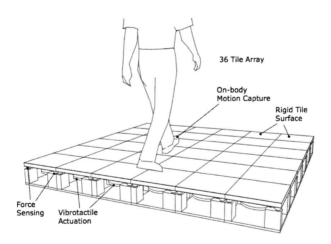

36 Tile Array

On-body
Motion Capture

Rigid Tile
Surface

Force
Sensing

Vibrotactile
Actuation

Fig. 4.3: Illustration of the distributed floor interface, with components labelled.

instrumented with four Interlink model 402 FSRs, which are located at the corners. Thus, the nominal linear sensing resolution of this array is 15 cm. In addition, each tile includes a wide bandwidth vibrotactile actuator [307], which although unused here, does modestly influence the sensor measurements, due to its weight [314]. Analog data from the force sensors are conditioned, amplified, and digitized via 32-

channel, 16-bit data acquisition boards. Each sensor is sampled at a rate of up to 1 kHz transmitted over a low-latency Ethernet link.

An array of six small-form-factor computers is used for force data processing.

For applications that do not require kinematic tracking, we infer foot-floor contact loci using intrinsic contact sensing techniques (see Chapter 3 and [314]), attaining an effective resolution of about 2 cm. However, such methods are incapable of tracking the location of a foot once it leaves the floor surface, and cannot resolve situations in which the feet overlap onto a single tile. Moreover, inference of lower body pose from in-floor force sensor data is challenging due to the loss of information inherent in this complex mapping. Such limitations have, in part, motivated the present work. In particular, we adopt the framework of Bayesian filtering in order to maintain continuity of lower body position estimates in dynamic settings, such as walking, where foot-ground contact is regularly interrupted.

Within this framework, our task remains challenging due to the high dimensional mapping from an articulated pedestrian pose to the observed force sensor values. In particular, this map is discontinuous for a sensing infrastructure consisting of independent, rigid tiles, as poses that are similar in nature, i.e. with footprints of similar 2D position and orientation, will lead to drastically different force observations when tile boundaries are crossed.

4.5.3 Tracking problem and algorithm

Consider a dynamic system with states x_t and observations y_t, both indexed by time. A Bayesian filter probabilistically estimate at time t the state x_t by sequentially updating a belief distribution $\mathrm{Bel}(x_t)$ over the state space, defined by $\mathrm{Bel}(x_t) = p(x_t|y_t, y_{t-1}, y_{t-2}, ...) = p(x_t|y_{1:t})$. Assuming the states comprise a Markov process, the Belief distribution at each subsequent time step can be obtained, using Bayes' Theorem, in terms of the belief state at the prior time step $t-1$, the assumed motion model $p(x_t|x_{t-1})$, and the likelihood $p(y_t|x_t)$ of the newly acquired observation y_t given the state x_t:

$$\mathrm{Bel}(x_t) \propto p(y_t|x_t) \int p(x_t|x_{t-1})\mathrm{Bel}(x_{t-1})\,dx_{t-1} \qquad (4.20)$$

Bayesian filters can be distinguished, in part, by the form of the Belief distribution $\mathrm{Bel}(x_t)$ that is assumed. Here, we adopt a Monte-Carlo approach, in which $\mathrm{Bel}(x_t)$ is represented by a set of weighted samples, or particles, given by $S_t = \{(x_t^i, w_t^i), i = 1, 2, ... N_s\}$. Our method uses the Sampling-Importance-Resampling (SIR) algorithm [97], described in Algorithm 1, and schematically illustrated in Figure 4.4.

For each new observation, SIR re-weighs the set of particles depending on their likelihood, evolves them in time using the assumed dynamic model, and updates the particle set S_t by resampling based on these weights.

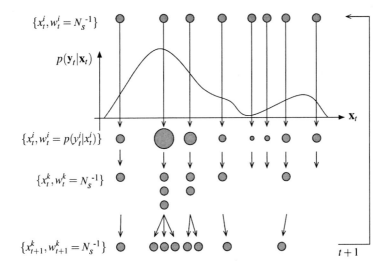

Fig. 4.4: The SIR particle filter algorithm.

Specifically, the likelihood function $L(x_t) = p(y_t|x_t)$ is computed in two steps as follows.

1. Using the observation model **H**, defined in Section 4.5.3.1, that maps states into the observation space, generate expected observations $y_t^* = \mathbf{H}(x_t)$.
2. Using the similarity measure $S(z, z') = p(z|z')$, defined in Section 4.5.3.2, compute the likelihood as $L(x_t) \equiv S(y_t^*, y_t)$.

As seen in (4.21), each particle is weighted according to its likelihood while also considering a prior distribution $p(x_t)$. As described in Section 4.5.3.2, the prior distribution $p(x_t)$ is useful for encoding constraints on states x_t, thus leading to an efficient search of the state space.

The attributed weight is then used to resample the particle set—see Equation (4.22). This step distinguishes SIR from SIS (Sampling Importance Sampling), and is meant to eliminate degeneracy of particles; by concentrating on high weighted particles to create a new uniformly distributed particle set, the extreme case where most particles have negligible weight is avoided. The opposite extreme, consisting of a single strong hypothesis, can be avoided by the inclusion of a roughening process. Roughening consists of the addition of random, zero-mean, normally distributed noise to all particles after resampling—see (4.24), to allow for a thorough search of the solution space. Additionally, particles are propagated forward in time based on the assumed dynamic model (4.23), which implicitly defines the motion probability $p(x_t|x_{t-1})$. Our dynamic model consists of two processes, accounting for the continuous and discrete aspects of the motion of interest. The dynamic model and roughening process are defined in detail in Section 4.5.4.

Algorithm 1 Sampling Importance Resampling

Initialize particles randomly: $S_0 \sim N(\mu_S, \sigma_S)$
while $t > 0$ **do**
    Observe y_t.
    for $i = 1$ to N_s **do**

$$\text{Likelihood:} \quad p(y_t|x_t^i) = S(y_t, H(x_t))$$
$$\text{Weight:} \quad w_t^i = p(x_t^i) \times p(y_t|x_t^i) \tag{4.21}$$

    end for
    for $i = 1$ to N_s **do**

$$\text{Normalize:} \quad W = \Sigma_i w_t^i, w_t^i \leftarrow W^{-1} w_t^i$$
$$\text{Resample:} \quad x_t^i \sim p(x_t^{i*}|w_t^i), w_{t+1}^i = N_s^{-1} \tag{4.22}$$
$$\text{Draw:} \quad x_{t+1}^i \sim p(x_{t+1}|x_t^i) \tag{4.23}$$
$$\text{Roughen:} \quad x_{t+1}^i \leftarrow x_{t+1}^i + \eta_{t+1}, \eta \sim N(0, \sigma_x) \tag{4.24}$$

    end for
    $S_{t+1} = \{(x_{t+1}^i, w_{t+1}^i)\}_{i=1}^{N_s}$
end while

In our system for tracking via in-floor force measurements, the relevant variables consist of:

- Observations y_t, consisting of a 12×12 array of force values, f_i.
- States, x_t, describing kinematic lower-body poses, are 19-dimensional vectors: $x_t = (\phi_{l,t}, \dot{\phi}_{l,t}, \dot{\phi}_{l,t-1}, \phi_{r,t}, \dot{\phi}_{r,t}, \dot{\phi}_{r,t-1}, \beta)$. They include planar midpoint coordinates u and orientations θ for each foot, where $\phi_l = (u_l, \theta_l)$ and likewise for ϕ_r, along with first time derivatives. The state x_t also includes a binary-valued vector $\beta = (\beta_l, \beta_r)$, implemented as a quaternary variable, which indicates the foot-floor contact condition ($\beta_i = 1$ if there is contact) for the left and right feet.

Figure 4.5 illustrates this state description within a skeletal model. The algorithm definition is completed by specifying the observation model $y_t = H(x_t)$, the likelihood model $p(y_t|x_t)$, and the motion model, $p(x_t|x_{t-1})$.

4.5.3.1 Observation model

We model expected observations $H(x_t)$ for a state x_t by simulating the mechanical forces associated with a pose. Ignoring shear forces, foot-tile contact results in a normal pressure distribution $p(u)$, where $u = (u, v)$ are 2D coordinates on the floor. The pressure distribution on a tile is conveniently summarized by a contact centroid $p_c = (u_c, F)$, where $F = \int du\, p(u)$ is the net normal force and u_c is the pressure centroid. A normal force with magnitude F, applied at u_c, would give rise to the same force measurements f_i as $p(u)$ [314, 36] (Figure 4.6).

Our observation model associates a pose x_t to a set of contact centroids, $p_{c,j}, j = 1, 2, \ldots, N_c$. One centroid is placed on each tile that a foot pose is determined to be

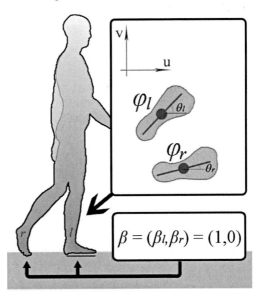

Fig. 4.5: State description of
lower body poses: $\mathbf{x}_t$ defines
feet and foot-floor contact.

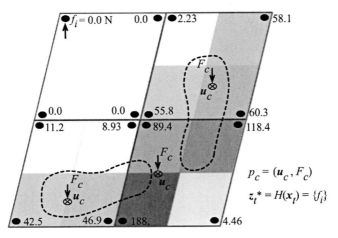

Fig. 4.6: The observation model $H(\mathbf{x}_t)$ maps a pose $\mathbf{x}_t$ (illustrated here as two feet) to a set of contact centroids ($\mathbf{u}_c$). These are then converted to force sensor readings $\mathbf{y}_t = \{f_i\}$. Expected force observations are illustrated as grids, with each quadrant intensity proportional to corresponding force sensor value (given in Newtons).

in contact with ($\beta_i > 0$), and the total weight F of the user is partitioned among these centroids. The placement of a contact centroid is determined by the average position of foot to tile contact. Expected sensor readings f_i are obtained from the static equilibrium equations for each tile. Each pressure centroid p_c yields a contribution $f_{i,c} = d_i^{-1} (\sum_{j=0}^4 d_j^{-1})^{-1}$, where $d_j = |\mathbf{u}_c - \mathbf{u}_j|$, and $\mathbf{u}_j$ is the sensor location. In this way, we predict observed force values $\mathbf{y}_t^* = \{f_{i,t}\}$ for a state $\mathbf{x}_t$.

4.5.3.2 Likelihood model

The likelihood function $L(\mathbf{x}_t) = p(\mathbf{y}_t|\mathbf{x}_t)$ describes the probability of observing force values $\mathbf{y}_t$ given the state $\mathbf{x}_t$. As described above, we use the observation model to generate expected observations, $\mathbf{y}_t^* = \mathbf{H}(\mathbf{x}_t)$. The likelihood function is then defined in terms of a normalized similarity measure $p(\mathbf{y}_t|\mathbf{x}_t) = S(\mathbf{y}_t^*, \mathbf{y}_t)$ between true force pattern observations $\mathbf{y}_t$ and expected force observations $\mathbf{z}_t^*$.

4.5.3.3 Similarity measure S

The similarity measure $S(\mathbf{z}, \mathbf{z}') = p(\mathbf{z}|\mathbf{z}')$ models the probability of observing $\mathbf{z}$ if the true observation is $\mathbf{z}'$. Conventional similarity measures make use of metrics such as Euclidean or Mahalanobis distance functions. However, since $\mathbf{H}$ is a high-dimensional discontinuous map from states $\mathbf{x}_t$ to observations $\mathbf{y}_t$, these measures cannot properly gauge similarity between observation vectors $\mathbf{y}_t$, $\mathbf{y}_t^* = \mathbf{H}(\mathbf{x}_t)$. More-over, the force observations are comparatively sparse, with most values being zero. As an alternative, we compute pair-wise similarity between such patterns, based on a measure of their area of overlap. Specifically, we employ a similarity measure that has proved useful in tracking via binary image masks [256], computing $S(\mathbf{z}^*, \mathbf{z})$ as the relative area of overlap between the true and expected 2D pressure distributions,

$$S(\mathbf{y}_t^*, \mathbf{y}_t) = \frac{\cap(\mathbf{y}_t^*, \mathbf{y}_t)}{\cup(\mathbf{y}_t^*, \mathbf{y}_t)} = \frac{1}{N_z} \sum_{i=1}^{N_z} \frac{\min(\mathbf{y}_t(i), \mathbf{y}_t^*(i))}{\max(\mathbf{y}_t(i), \mathbf{y}_t^*(i))}$$

Figure 4.7 illustrates conceptual examples of observations $\mathbf{y}_t$, $\mathbf{y}_t^*$ as 2D pressure distributions, as well as their intersection $\cap(\mathbf{y}_t^*, \mathbf{y}_t)$ and union $\cup(\mathbf{y}_t^*, \mathbf{y}_t)$.

We note that the average overlap of these resulting pressure distributions is an effective metric for capturing similarities between force observations as a probability:

$$\begin{aligned}
S(\mathbf{z}, \mathbf{z}) &= 1 \\
\mathbf{y}_1 \neq \mathbf{y}_2 &\Rightarrow 0 < S(\mathbf{y}_1, \mathbf{y}_2) < 1 \\
S(\mathbf{y}_1, \mathbf{y}_2) &= S(\mathbf{y}_2, \mathbf{y}_1)
\end{aligned} \qquad (4.25)$$

4.5.3.4 Postural constraints

The likelihood model is modified to encode human postural constraints, via a prior distribution $p(\mathbf{x}_t)$. The latter is defined to consist of a set of independent postural priors for human walking, in the form of univariate Gaussian distributions $N(\mu_\varsigma, \sigma_\varsigma)$, $N(\mu_\Theta, \sigma_\Theta)$ over stance width $\varsigma = \|\mathbf{u}_l - \mathbf{u}_r\|$ and relative orientation $\Theta = |\theta_l - \theta_r|$ respectively. The postural prior is introduced as the product $p(\mathbf{x}_t) = N(\varsigma; \mu_\varsigma, \sigma_\varsigma) N(\Theta; \mu_\Theta, \sigma_\Theta)$. The prior distribution $p(\mathbf{x}_t)$ is applied when computing the particle weights from the likelihood L—see (4.21).

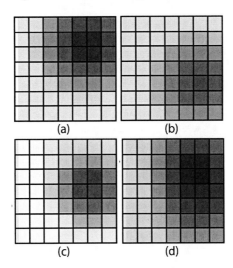

Fig. 4.7: Computations of the similarity measure from the 2D pressure distributions for: (a) $\mathbf{y}_t$, (b) $\mathbf{y}_t^*$, (c) $\mathbf{y}_t \cap \mathbf{y}_t^*$, (d) $\mathbf{y}_t \cup \mathbf{y}_t^*$. Pressure distributions are illustrated as grids, with each quadrant intensity proportional to corresponding force sensor value. The intersection ($\cap$) and union ($\cup$) of force observations are used in computing similarity $S(\mathbf{y}_t, \mathbf{y}_t^*)$.

4.5.4 Dynamics model

We model the movements of the lower body of a walker via the motion probability $p(\mathbf{x}_t | \mathbf{x}_{t-1})$. The state x_t consists of continuous configuration variables ϕ_i and discrete contact variables β_i. We therefore approximate foot motion in a hybrid (stochastic) framework, incorporating continuous, linear movements of the limb coupled to discrete state transitions reflecting changes of the foot-floor contact conditions.

4.5.4.1 Continuous, linear dynamics

The state representing the left or right foot is denoted respectively by a vector $\phi_i = (\mathbf{u}_i, \theta_i)$ giving the midpoint and orientation of either the left ($i = l$) or right ($i = r$) foot at time t (Figure 4.5). The dynamics used in our algorithm can be described by the following linear, discrete time system:

$$\phi_{i,t} = \phi_{i,t-1} + \beta_i \Gamma_i(\dot{\phi}_{t-1} dt + \eta_t^0) \tag{4.26}$$

$$\dot{\phi}_{i,t} = \beta_i[\alpha \dot{\phi}_{i,t-1} + (1-\alpha)(\dot{\phi}_{i,t-2} + \eta_{t-1}^1)] \tag{4.27}$$

The parameter β_i is the binary contact variable for foot i. Thus, ϕ_i is constant when there is contact ($\beta_i = 0$) and otherwise drifts, with position and velocity driven by additive Gaussian noise processes η_t^0 or η_t^1, where $\eta \sim N(0, \Sigma)$. For efficiency, we parametrize drift via a single noise process, defining $\eta \equiv \eta_t^0 = \eta_t^1 dt$ for all t. The noise covariance Σ is a 3×3 diagonal matrix with diagonal entries $\sigma_u, \sigma_v, \sigma_\theta$. To mimic walking, velocity drift in the direction that the foot is oriented is assumed to be larger. This non-isotropic drift is implemented through the factor Γ_i, a diagonal matrix with entries $(\cos\theta_i(\gamma + \sqrt{1-\gamma^2}), \sin\theta_i(\gamma - \sqrt{1-\gamma^2}), 1)$, where $0 < \gamma < 1$ is a dimensionless scalar defining the longitudinal bias. α is a dimensionless scalar defining the velocity noise mixing rate. It approximates the dynamics of a free foot during walking by means of a saturating linear drift velocity.

Although this continuous dynamic model violates the Markov assumption made in Equation (4.20), we note that there exists a Markov representation of this system, via the following change of variables:

$$\Phi_t = \begin{bmatrix} \dot{\phi}_t \\ \dot{\phi}_{t-1} \end{bmatrix} = \begin{bmatrix} \alpha & 1-\alpha \\ 1 & 0 \end{bmatrix} \Phi_{t-1} \qquad (4.28)$$

4.5.4.2 Roughening

As presented in Algorithm 1, noise $\eta \sim N(0, \sigma_x)$ is added to the continuous state components ϕ_i at each SIR step in order to avoid particle degeneracy.

4.5.4.3 Discrete state transition model

The dynamic model includes discrete transitions from the foot-floor contact states $\beta = (\beta_l, \beta_r)$, where $\beta = 0$ or 1, via the stochastic process, shown in Figure 4.8. Despite its simplicity, this model is effective in approximating discrete stepping

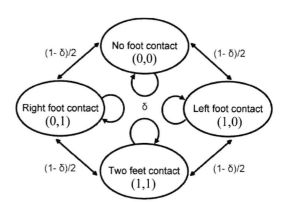

Fig. 4.8: Stochastic state transition diagram approximating stepping motion.

motion. δ is an empirically determined probability of no change in contact state β. All remaining transitions are symmetric, with transition probability $\frac{1-\delta}{2}$.

4.5.5 Experiment and results

The system described above was evaluated by measuring the absolute positions of the feet of pedestrians using data acquired synchronously via motion capture (Vicon Motion Systems). Reflective markers were attached to the walkers' shoes, providing an accurate estimate of 3D foot positions. Five recordings of walking sequences between 5.7 and 12.4 seconds in length were acquired via the apparatus described in previous reports. Synchronous motion capture and force data were recorded. Errors were computed based on maximum a posteriori (MAP) foot position estimates obtained from the tracking algorithm.

Figure 4.9 shows the state estimates and force observations at four stages of the walking sequence: the initial particle set has high variance, but is quickly narrowed down to a few hypotheses which are evolved based on our motion model. Figure

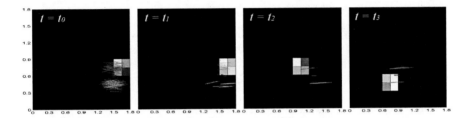

Fig. 4.9: Force observations and pose estimates at 4 stages of the sample walking sequence. Observations are shown as 2D pressure distributions, with quadrant intensity proportional to force sensor readings. Poses are illustrated as green and red lines, corresponding to a top view of left and right feet estimates respectively, with blue circles corresponding to inferred center of mass.

4.10 shows the resulting motion trajectories and errors for the two foot locations, in planar coordinates. Average RMS error values are reported in Table 4.1.

Foot	Error (m)	Windowed error (m)
Right (no contact)	0.1901	0.1245
Left (no contact)	0.1649	0.0336
Average (no contact)	**0.1775**	**0.0791**
Right (contact)	0.2025	0.1058
Left (contact)	0.1406	0.0306
Average (contact)	**0.1716**	**0.0682**
Right (all)	0.1982	0.1122
Left (all)	0.1490	0.0316
Average (all)	**0.1736**	**0.0719**

Table 4.1: RMS Position Error

The experimental parameters are given in Table 4.2. Position errors during foot-floor contact are found to be slightly less on average than when a foot is not

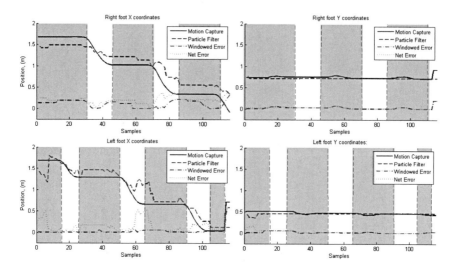

Fig. 4.10: Sample results for a pedestrian crossing the apparatus, comparing measurements from motion capture data and estimates obtained from the particle filter. Results are shown for both feet ϕ_l, ϕ_r. The shaded domains illustrate contact states, with grey and white corresponding to $\beta_i = 1$, $\beta_i = 0$ respectively.

SIR Algorithm	
Number of particles	$N_s = 200$
Initial Noise	$\sigma_S = 4.572$ cm
Roughening Position Noise	$\sigma_x = 0.003048$ cm/rad
Likelihood Model	
Postural prior stance size	$(\mu_\varsigma, \sigma_\varsigma) = (30.48, 30.48)$ cm
Postural prior stance angle	$(\mu_\Theta, \sigma_\Theta) = (0, \pi/2)$ rad
Dynamic Model	
Additive Position Noise	$\sigma_{u,v} = 0.3048$ cm
Additive Angle Noise	$\sigma_\theta = 0.003048$ rad
Longitudinal bias	$\gamma = 0.9$
Velocity mixing rate	$\alpha = 0.55$
Probability of no β change	$\delta = 0.55$

Table 4.2: Experimental parameters

in contact with the floor. Temporal alignment mismatches were found to have a large effect, so we also performed a windowed error calculation in which an acceptable time shift of 10 samples (at 20 Hz) was permitted. This greatly reduced RMS position errors (see Table 4.1), suggesting that system tracking performance may be most acceptable in situations in which temporal accuracy is not important. Tracking performance in more temporally demanding settings might be greatly improved if a better alignment can be achieved. Video documentation of these results is provided in the supplementary material, and available online at http://www.cim.mcgill.ca/~rishi/video.swf.

In addition to position estimates, this system provides continuous labels identifying the walker's right and left feet. During tracking of the sequences used for evaluation, left and right feet were continuously and coherently identified with 100% accuracy. The capability of this system to maintain and propagate these labels may be useful for applications including floor-based touch screen user interfaces, where it may be desirable to assign each foot a different functional operation, or to render a different response to left and right foot [314].

4.5.6 Person tracking from in-floor force measurements: Conclusions

This section described the application of Bayesian filtering techniques to the problem of tracking the lower body pose of a pedestrian from foot-floor interaction forces acquired via a coarse array of in-floor force sensors. The system developed achieves continuous and labeled tracking of the lower limbs of a walker via a coarse sensor array, by combining prior knowledge about the mechanical structure of the interface and a simple, but consistent, model of the dynamics of the feet during walking. In the experiment described above, our system never confused the left and right feet of the walker, and was able to track the locations of each with an average resolution on the order of 15 cm, and with improved resolution when feet are in contact with the surface. Potential applications of these techniques include tracking in smart environments in which motion capture is impractical (due to occlusion or other factors), and to interaction with distributed, floor-based touch surface interfaces for the feet [314, 313].

Despite the promising nature of these results further experiments are needed to evaluate the quality of the tracking. In addition, it is clear that the system itself can be improved in several respects. A higher-density sensor network would improve position estimates during contact, albeit at greater cost. A model for the non-contact portion of walking movement that is more sophisticated than the random drift model used in our system could significantly improve estimates in foot tracking when a foot is not in contact with the floor. The incorporation of additional prior knowledge about the kinematic constraints on lower limb positions during walking would also be expected to contribute improvements. In ongoing work, we are exploring the possibilities for optimally fusing information from in-floor force sensors with motion capture or other video sensors, in order to resolve data loss due to camera occlusion, or to provide contact forces and timing information that cannot be accurately estimated from video.

4.6 Conclusion

This chapter discussed aspects of the problem of tracking of pedestrian movements in augmented environments, reviewing the selection of ambient sensing modalities and devices, the extraction of information from walking-generated sensor data, and a selection of the wide variety of algorithms that have been used for mining the resulting data and for tracking the movements of persons in space. In the last section, a sample application was described based on tracking pedestrians from in-floor force sensor measurements, using an interface that is described more fully in another chapter of this book.

Chapter 5
Novel interactive techniques for walking in virtual reality

L. Terziman, G. Cirio, M. Marchal, and A. Lécuyer

Abstract In this chapter we present a state of the art of interactive virtual walking in bounded physical spaces. We will focus on three novel approaches achieving, in different ways, the challenging objective of affording interactive navigation in large virtual environments with limited workspaces. The underlying concepts stem from a development roadmap aiming at progressively empowering gestural control as a mean for virtual locomotion, furthermore moving from hardware issues until touching questions that are more related to software.

5.1 Introduction

Navigation is a fundamental interaction task in virtual environments. Most VR applications in fact give users the possibility to walk and/or move in the virtual world.

One constraint that often comes along with VR setups is given by the limited workspace in which users are physically walking. Their motions in fact are bounded by either the walls of the simulation room or the range of the tracking system. Virtual navigation techniques must therefore cope with such workspace restrictions.

In this chapter we present three novel approaches to virtual navigation which achieve, in different ways, the challenging objective of interactive virtual navigation in large virtual environments with limited workspace. The concepts behind afford progressively more effective physical walking, meanwhile gradually shifting the design focus from hardware to software issues.

- The first approach results in an input device called *JoyMan*, which can also be seen as a sort of "human-scale joystick". It is a novel piece of hardware exploiting human *equilibrioception*: users can control their virtual navigation speed and orientation by just leaning with their body.
- The second approach, called *Shake-Your-Head*, is more linked to the walking act since users can *walk in place* to control a potentially infinitely long virtual

walk. Furthemore, Shake-Your-Head makes use of a simple webcam to sense head motion during walking in place.

- The third approach, called *Magic Barrier Tape*, is a virtual interactive signal warning about the actual workspace limits. This concept enables safer and more effective physical walking across the virtual world. Users can push against the virtual tape to switch the navigation mode to *rate* control, and afford otherwise unreachable locations.

In the remainder of this chapter we will first review related work in the field of walking and navigation in VR. Then, we will describe the above three approaches in detail. Finally, we will discuss and summarize the relative pros and cons of each of them in ways to help interaction designers make the most appropriate choice, also depending on their application needs.

5.2 Virtual walking in bounded workspaces

There is an important body of work on human navigation in VR. We first review techniques which allow navigation with neither physical walking nor spatial movement of the user. In other words they afford static navigation and walking in place. We then discuss navigation techniques involving physical walking, which are more relevant to our work.

5.2.1 Static navigation and walking in place

There are several known VR locomotion metaphors in which the user is not required to walk [43], and therefore does not need to deal with workspace restrictions. Examples of these metaphors include "teleportation", that is an instantaneous switch to a new location. Worlds In Miniature (WIM) [231] is a metaphor in which users hold a copy of the virtual world in their hands; from that copy, they can point to a location and be brought anywhere in the virtual world. Probably the most common navigation technique is the Flying Vehicle, where the environment is not manipulated; the illusion is that the user can move through the world, either by using a mock-up, a wand or other device.

Walking in place [273, 293], shortly WIP, simulates the physical act of walking without forward motion of the body; a virtual forward motion is introduced instead. The optical flow, that should match with the proprioceptive information coming during the physical walking act, is instead coupled to virtual proprioceptive cues. The sense of presence is greatly increased compared to static navigation techniques [304], though, other (mainly the vestibular) sensory cues of walking are missing.

5.2.2 Physical walking

Several studies have shown the benefits of affording physical walking during the navigation of virtual environments, in terms of task performance [261, 116, 332] as well as sense of presence [304] and naturalness [152, 304, 332]. Hence, several techniques allow users to walk inside bounded workspaces.

In Step WIM [157] users invoke a miniature version of the world at their feet, allowing to navigate across proportionally greater distances. Over this miniature, users can walk to a new location of the WIM and then trigger a command that rescales it until reaching the size of the virtual world. When used inside a CAVE, where the visual field is limited by the missing screen, a 360° map of the scene is displayed. The resulting effect requires some adaptation time to users, and does not entirely solve the problem coming from the missing screen.

The Seven League Boots [133] allow to walk in virtual worlds that are larger than the real space. They scale the users' motion speed only along the traveling direction, using previously acquired information on gaze and motion. Although appreciated, it does not entirely solve the limited workspace problem.

Resetting techniques [327, 328] try to overcome the workspace limitations through the use of HMDs. These techniques reset the user's position or orientation in the real world when reaching the workspace boundaries, without breaking the spatial awareness of the virtual world. In the Freeze-backup technique, the virtual world is "frozen" and the user steps back until reaching the workspace centre. In the *Freeze-turn* technique, the users' orientation is frozen while they physically pivot by 180°. In the *2:1-turn* technique, a 360° virtual rotation is mapped to a 180° real world rotation, and the user also physically pivots by 180°. These resetting techniques are performed consciously by the users, after they are warned by a specific signal which implies a break in immersion. Moreover, the resetting action itself might be felt as unnatural.

5.2.3 Redirection techniques

Redirected walking [253, 278, 149] and Motion compression [205, 89, 281] techniques let users walk along a curved path in the real world when following a virtual straight line, through the progressive rotation of the scene. In a sufficiently large workspace the user can walk endlessly without reaching the boundaries of the real workspace. These techniques in some cases are perceptually transparent, however they require large workspaces and can create confusion in users when they make unpredictable or quick changes of direction. Redirected walking is more effective if implemented using HMDs and wide area tracking systems; confusion can be relieved by introducing distracting events [233].

Redirected walking was used in combination with a walking in place technique in a CAVE environment, to smooth the missing screen problem [254]. Experiments showed that the frequency of the looks at the missing screen was not reduced compared to hand-held navigation techniques, although the variance was. In the specific case of architectural virtual walkthroughs, the Arch-explore interface [49] allows the use of redirection techniques in small environments such as CAVEs.

Change blindness redirection [284] redirects users wearing HMDs by making dynamic changes to the environment, such as shifting the door positions and re-configuring corridors while the user is focusing on a distracting task. This recent technique is highly transparent, hence very promising.

In the conclusion of this section, we can say that there is lack of adequate navigation techniques allowing immersive real walking and infinite exploration for CAVE-like environments. Existing techniques exhibit limitations, and there is no general solution to the missing screen problem. Although promising, redirection techniques require specific real and virtual environments that let them work correctly. In the following we will address these issues, by proposing three novel navigation techniques for safe, natural and enjoyable navigation within CAVE and similar simulation environments.

5.3 JoyMan: Navigating in virtual reality using equilibrioception as a human joystick

JoyMan is a new interface for navigation in virtual worlds. The main objective of this interface is to afford realistic locomotion trajectories in the VE. JoyMan meets this objective by combining two components: a peripheral device allowing users to select the desired direction of locomotion, and a control law which maps the device state into a virtual velocity vector.

5.3.1 Description of the JoyMan interface

The mechanical design of the device is inspired by the metaphor of a human-scale joystick. It mainly consists of a board upon which the user stands, and whose inclination depends on the user position. Users point to a locomotion direction by tilting the platform toward the target.

The mechanical design of the platform allows users to change the inclination by leaning, meanwhile preventing them from falling for a reactive mechanical force constantly pulls the platform toward the horizontal position. A control law maps the angles of pitch and roll (forming the device *state*) into a velocity vector informing the virtual locomotion. The proposed law ensures realistic virtual walking: particularly, the tangential velocity is bounded and depends on the angular velocity, as detailed in the following.

5.3.1.1 Device mechanical design

The mechanical architecture of the interface is illustrated in Figure 5.1. The key

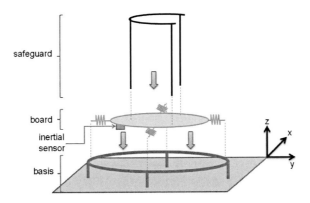

Fig. 5.1: Schematic representation of the main elements forming JoyMan. The board and the base are connected through springs, generating a reactive force that recalls the board to the horizontal plane. The board can be oriented around the two horizontal axes (pitch and roll) with limited range. Users act on the device by standing on it, and tilt the board by leaning with their body. The board inclination is measured using an inertial sensor. The safeguard prevents users from falling when they lean.

principle of JoyMan is to let the user point to the desired direction of virtual locomotion by leaning in the corresponding direction. In other words, locomotion is controlled by *equilibrioception*.

JoyMan features the following functionalities:

- users should be able to stand on the platform, and to assume a natural posture furthermore compatible with the locomotion task;
- users should feel perfectly safe when using the platform which, in parallel, must prevent them from falling;
- they should be able to lean beyond their own equilibrium limit, in order to increase the magnitude of their vestibular sensations;
- the device should keep the user in vertical position to avoid fatigue during use.

Other features were also considered during the design of the device: the most important one was to come up with an affordable interface. Exploring equilibrioception in place of proprioception to control one's locomotion in the virtual world is also a promising direction to decrease simulation costs. JoyMan does not make use of active mechanical devices, nor implies the use of sophisticated tracking systems. The objectives listed above are based on a mechanical architecture, as simple as that illustrated in Figure 5.1.

The device is made of 4 components:

The *base*   supports the whole architecture. It is made of a flat square platform lying on the ground, along with a circular metal frame linked to the board by legs. The legs and the platform are welded together and the board is large enough to prevent the device from tipping over.

The *board*   supports users. It is circular, and linked to the frame through springs. As a result, the board floats free within a limited range along two degrees of freedom, that are the rotations around the x and y axis as illustrated in Figure 5.1.

The *safeguard*   is rigidly fixed to the board and prevents users from falling.

The *inertial sensor*   is rigidly attached to the board, and records its current orientation.

The board is made of a square woodcut platform. The metal frame (1 m diameter) and the springs came from an old trampoline. The base consists of a plastic plate having a diameter of 0.55 m, equipped with 18 hooks on its rim. A rigid rope is tied up between the hooks, and the springs attach the board to the frame. Finally, the safeguard is made of welded iron tubes. Its height, ranging between 0.8 m and 1.2 m, can be adapted to users. Costs for its prototyping did not exceed $500, excluding the inertial sensor.

Some illustrations of JoyMan in action are provided in Figure 5.2. The user, standing on the platform, must initially lean to start the virtual locomotion. The control law that is employed during the navigation is described in the next section.

5.3.1.2 Virtual locomotion control

The main steps leading to the virtual locomotion control are summarized in Figure 5.3. The control is composed of two components: the locomotion model, and the control law. Both are detailed in the following paragraphs.

Locomotion model

The locomotion model allows to translate the position and orientation of the platform into virtual motion in the VE. Some experimental observations on the human walking trajectory have been made prior to building our model [12].

The position P of the user in the virtual space can be defined as an oriented point moving in the horizontal plane (see Figure 5.3):

$$P = \begin{bmatrix} x \\ y \\ \theta \end{bmatrix}.$$ (5.1)

The virtual motion is controlled by the velocity. We assume the virtual trajectory is non-holonomic, meaning that the velocity vector orientation and the body orientation are always identical. The non-holonomic nature of walking human trajectory has been experimentally observed in [12].

Non-holonomy allows to decompose the velocity vector V as follows:

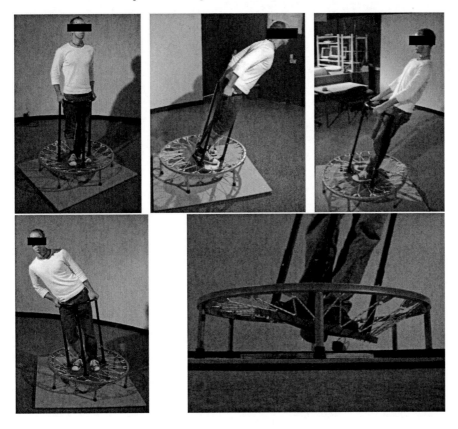

Fig. 5.2: Illustrations of the device in use. Users can stand straight on the board, reactive forces make interaction easy. Users have to lean to start locomotion in virtual worlds. All possible leaning directions are displayed.

$$V = v_t \cdot \begin{bmatrix} \cos \theta \\ \sin \theta \end{bmatrix}. \tag{5.2}$$

Such a decomposition allows to independently control the tangential speed v_t and the orientation θ. Human tangential velocity is limited. We denote with $v_{t_{max}}$ the maximum tangential velocity bound. By default, we set this bound to $v_{t_{max}} = 1.4$ m/s. It has equally been observed that very slow walking velocities are never reached: such velocities are humanly feasible but not used in practice. We thus define $v_{t_{min}} = 0.6$ m/s. However, it has been observed that changing orientation θ affects the amplitude of v_t during human locomotion [334]: humans decelerate when turning. In order to take into account such an observation, we define $v_{t_{max}}^{\dot{\theta}}$ to be the bound of the tangential velocity once knowing the current turning velocity $\dot{\theta}$. This bound is defined as follows:

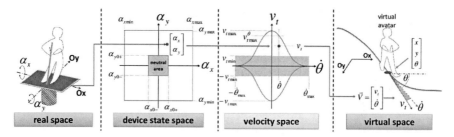

Fig. 5.3: Summary of the different steps for the virtual locomotion control. From the position and orientation of JoyMan, we can compute the virtual velocity vector by using a locomotion model and control law.

$$v_{t_{max}}^{\dot{\theta}} = a \cdot v_{t_{max}} \cdot e^{-b \cdot \frac{\dot{\theta}}{c}}$$
$$v_{t_{min}}^{\dot{\theta}} = a \cdot v_{t_{min}} \cdot e^{-b \cdot \frac{\dot{\theta}}{c}}$$

(5.3)

where a, b and c are parameters. As a result, we model the reachable tangential velocity $v_{t_{max}}^{\dot{\theta}}$ as a Gaussian function of the current turning velocity $\dot{\theta}$. The higher $\dot{\theta}$, the lower the tangential velocity bound. By default, we arbitrarily choose: $a = 1.07$, $b = 0.5$ and $c = 0.7$. Such values match the experimental observations provided in [334]. Finally, the absolute value of the angular velocity is also bounded to $\dot{\theta}_{max}$. We arbitrarily choose $\dot{\theta}_{max} = 1$ rad/s.

Control Law

The control law allows users to modify the virtual velocity vector V by leaning on the device. The modification of the platform orientation affects the state of the device s_d. The value of s_d is defined by the orientation α_x and α_y of the board relatively to the two horizontal axis x and y, as measured by the inertial sensor:

$$s_d = \begin{bmatrix} \alpha_x \\ \alpha_y \end{bmatrix}.$$

(5.4)

We neglect the orientation around the z axis. During calibration, we ask the user to stand on the platform and then to firmly lean towards all the cardinal directions. We estimate the reachable bounds of the board orientation by averaging these bounds over a short period of time, $\alpha_{x_{min}}$, $\alpha_{x_{max}}$, $\alpha_{y_{min}}$ and $\alpha_{y_{max}}$. We also ask the user to stand straight on the platform and define a neutral area bounded by $\alpha_{x_{0+}}$, $\alpha_{x_{0-}}$, $\alpha_{y_{0+}}$ and $\alpha_{y_{0-}}$.

We want the user to control her tangential velocity v_t by leaning forward or backward, i.e. by playing with α_x, whereas the angular velocity $\dot{\theta}$ is controlled by leaning aside, i.e., by playing with α_y. The angular velocity is controlled as follows:

$$\begin{cases} \dot{\theta} = \dot{\theta}_{max} \cdot \frac{\alpha_{ymax} - \alpha_y}{\alpha_{ymax} - \alpha_{y0+}} & \text{if } \alpha_y > \alpha_{y0+} \\ \dot{\theta} = \dot{\theta}_{max} \cdot \frac{\alpha_y - \alpha_{ymin}}{\alpha_{y0-} - \alpha_{ymin}} & \text{if } \alpha_y < \alpha_{y0-} \\ \dot{\theta} = 0 & \text{otherwise} \end{cases} \tag{5.5}$$

which allows to infer $v^{\dot{\theta}}_{t_{max}}$ and $v^{\dot{\theta}}_{t_{min}}$ according to (5.3), and finally v_t:

$$\begin{cases} v_t = v^{\dot{\theta}}_{t_{min}} + (v^{\dot{\theta}}_{t_{max}} - v^{\dot{\theta}}_{t_{min}}) \cdot \frac{\alpha_{xmax} - \alpha_x}{\alpha_{xmax} - \alpha_{x0+}} & \text{if } \alpha_x > \alpha_{x0+} \\ v_t = -v^{\dot{\theta}}_{t_{min}} + (v^{\dot{\theta}}_{t_{max}} - v^{\dot{\theta}}_{t_{min}}) \cdot \frac{\alpha_x - \alpha_{xmin}}{\alpha_{x0-} - \alpha_{xmin}} & \text{if } \alpha_x < \alpha_{x0-} \\ v_t = 0 & \text{otherwise} \end{cases} \tag{5.6}$$

At each temporal step n, the virtual velocity vector $V_n = [v_{t_n}, \dot{\theta}_n]$ is thus calculated from s_d, the state of the device. Before updating the simulation accordingly, we check that no unrealistic acceleration is performed. Thus, given the previous velocity vector V_{n-1}, we finally compute the current velocity vector V_f:

$$V_f = V_{n-1} + \lfloor \frac{V_n - V_{n-1}}{\Delta t} \rceil \tag{5.7}$$

where Δt is the simulation time step, and $\lfloor \rceil$ denotes truncation in a way that the absolute tangential acceleration does not exceed 1 m/s^{-2} and the angular acceleration does not exceed 1 rad/s^{-2}.

5.3.1.3 JoyMan interface

The JoyMan interface is designed for immersive virtual locomotion into virtual worlds. It is mainly aimed at exploring the possibility of equilibrioception in place of proprioception, in contrast to many current interfaces. The mechanical design of this platform is relatively simple. We showed how the proposed device allows a user to safely perform exaggerated leaning motion, beyond the normal equilibrium limit, in order to control virtual navigation. Here we remind our long-term objective, that is, performing realistic locomotion in the virtual world. What can we expect from the JoyMan interface?

Immersion

The device does not pose technical limits to visual or audio feedback design choices. JoyMan can be used virtually in any immersive environment. Users signal the beginning of a locomotion task by leaning ahead: this act has similarities with the kinematics of walking, which involves continuous unbalancing along the direction of locomotion. We expect that this similarity, preserving some vestibular cues of walking, has potential to improve immersion.

Realism

The dynamics of some user's actions needed to operate JoyMan resembles real walking motion. As an example, users perform a left turn followed by a right turn by reversing the platform orientation using the whole body. Involved inertias and frictions prevent from performing this change of state immediately: this happens also during walking, when the inclination of the body is used to counterbalance momentum. In spite of these similarities, we do not expect users to achieve efficient navigation (e.g. in terms of task completion time) using JoyMan.

5.3.2 Preliminary evaluation

As a preliminary evaluation, JoyMan was compared against the joystick, a traditional interface that is still regularly used in VR applications. The joystick is one of the most performing navigation interfaces in terms of task completion time.

The first goal was to quantify the relative loss of performance. The experiments were conducted by displaying a 3D VE either on a screen or on a HMD. The effectiveness of the interface was investigated in a locomotion task across complex pathways, composed of different gates placed in the VE.

The second goal was to compare the sense of immersion between the two interfaces. A subjective questionnaire was proposed to the participants, to evaluate their own preferences in terms of quality of the navigation in the VE.

5.3.2.1 Performances: Task completion time and accuracy

For each participant, the task completion time of each trial was measured for the different experimental conditions. A repeated two-way ANOVA was performed on the two different interfaces and the two visual conditions. The ANOVA accounting for the visual conditions and the task completion time revealed a significant effect ($F(1,526) = 11.63$, p-value< 0.001). A significant effect was also found for the interfaces and the task completion time ($F(1,526) = 54.47$, p-value< 0.001). As expected, the results reveal that the joystick was better than JoyMan in terms of speed of the navigation. The mean value for the completion time of a path was 187 s (std = 13 s) for the joystick and 321 s (std = 89 s) for JoyMan.

A specific analysis was developed to study learning effects using both interfaces. A linear model, where all conditions were mixed, was fitted to explain the relation between the task completion times and the trial number. For the JoyMan configuration, it revealed that the slope of the linear regression was significantly lower than zero (p-value < 0.000001), reflecting a significant decrease in the task completion time as the number of trials increases. However the same analysis, in which the first trial was removed, showed that the slope was not significantly different from zero

(for both visual conditions). Figure 5.4 illustrates the task completion time of the
different trials for the JoyMan configuration.

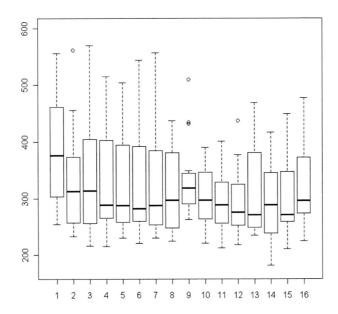

Fig. 5.4: Task completion time (in seconds) for the JoyMan configuration (16 trials). The first trial
is included to illustrate the learning effect. Each box plot is delimited by the quartile (25% quantile
and 75% quantile) of the distribution of the condition over the individuals. The median is also
represented for each trial.

For each participant and for each trial, the percentage of errors for the different
paths was measured. There was no error at the end, for both configurations (JoyMan
and joystick) and both visual conditions.

5.3.2.2 Subjective questionnaire

A preference questionnaire was proposed in which participants had to score from
1 (low appreciation) to 7 (high appreciation) the two configurations (JoyMan and
joystick) according to eight subjective criteria: Fun, Intuitive, Accuracy, Presence,
Rotation realism, Fatigue, Cybersickness and Global appreciation. The grade 7 for
Fatigue and Cybersickness respectively means that the interface did not induce any
fatigue and did not imply any cybersickness feeling.

After performing an ANOVA on the two different conditions, a significant effect for six criteria was found: Fun ($F(1,30) = 17.77$, p-value < 0.001), Intuitive ($F(1,30) = 21.25$, p-value < 0.001), Accuracy ($F(1,30) = 23.52$, p-value < 0.001), Presence ($F(1,30) = 13.35$, p-value < 0.001), Rotation realism ($F(1,30) = 6.63$, p-value $= 0.015$) and Fatigue ($F(1,30) = 87.51$, p-value < 0.001). In particular, JoyMan was better ranked concerning Fun, Presence and Rotation realism. No significant effect was found for Cybersickness ($F(1,30) = 4.01$, p-value $= 0.054$) and Global appreciation ($F(1,30) = 0.69$, p-value $= 0.411$).

5.3.3 Discussion and perspectives

As a global conclusion of the preliminary evaluation of JoyMan, it can be claimed that the feeling of immersion in the virtual world is significantly improved compared to traditional joystick-based techniques, at the cost of some difficulty of use. JoyMan is still at an early stage of development. However, the preliminary results are promising, and open a large set of possible directions to improve usability and the level of realism of virtual navigation. This section discusses these directions as thoroughly as possible, as well as future work to meet our objectives: an easy, intuitive, immersive interface allowing realistic locomotion in virtual worlds.

5.3.3.1 Interface calibration

The navigation with JoyMan involves the whole body and, as expected, the task completion times were higher for this interface compared to the joystick (where only the arms and hands are involved). The use of the whole body, implying more movements, is also a reason for the lower rating for the Accuracy criterion in the subjective questionnaire. However, the results concerning the evaluation of JoyMan in terms of performances are encouraging as the participants always succeeded to complete the navigation task during the evaluation. One possibility for improving the performances of JoyMan may consist of a better interface calibration. The control law directly transforms the inclination of the board into virtual walking velocities. Three types of parameters were proposed to design the control law: those defining the active angles area into the device state space ($\alpha_{x_{min}}$, $\alpha_{x_{max}}$, $\alpha_{y_{min}}$ and $\alpha_{y_{max}}$, $\alpha_{x_{0+}}$, $\alpha_{x_{0-}}$, $\alpha_{y_{0+}}$ and $\alpha_{y_{0-}}$), those controlling the reachable virtual velocities (θ_{max}, $v_{t_{min}}$ and $v_{t_{max}}$) and finally those controlling the dynamics of the relation between tangential and angular velocities a, b, and c as by (5.3).

Ideally, the first type of parameters should be calibrated. In fact, users took some effort to reach a given inclination (corresponding to a certain walking velocity) depending on their size, weight and strength. The fatigue reported by participants in the experiments confirms the need for individual calibration. Currently, the calibration process consists of recording leaning motions as well as neutral positions, to define the boundaries of the active area of the platform. This process was carried out before the experiments, however in the end one single setting was kept in all evaluations hence locking the behavior of the interface irrespectively of the participants. For VR applications, we envisage to explore various calibration techniques as well, such as displaying a VE with a moving point of view, or asking users to subjectively apply an effort onto the platform corresponding to the expected amount of motion.

5.3.3.2 Mechanical design

The results of the subjective questionnaire suggest that the participants enjoyed the navigation with JoyMan. They namely gave a higher rating for the Fun criterion, but also for Presence and Rotation Realism. Specifically, the higher rate given to the last criterion confirms that the mechanical design was appreciated for locomotion tasks in the VE. Improvements may concern the links between the platform and the base. The proposed prototype in fact implements a simple connection, based on an inextensible rope and a set of springs (see Figure 5.2). This connection may be conveniently modified, to add flexibility of movement to the interface. Mechanically speaking, the link could ideally realize two rotational degrees of freedom, one with a restoring force proportional to the platform inclination.

5.3.3.3 Control law

The experimental evaluation reveals that the proposed control law is intuitively grasped by users: establishing a relation between linear and angular velocities seems to be naturally accepted, and is consistent with observations of human locomotion trajectories. Future work will deal with modifications of the control law, for example aiming at lowering vestibular and visual sensory conflicts.

The device state vector is currently two dimensional, and describes the two angles of platform inclination. Experiences showed that JoyMan can be fully and accurately controlled by the lower body movements only; it does not even require to handle the barrier (except for moving backward in the current state of the device). Such a feature opens interesting perspectives, and makes possible to increase the dimension of the device state space. Possible extensions are numerous, and immediately within reach if the VR system will be equipped with tracking abilities: hands may remain free to achieve secondary actions in the virtual worlds (grasping, touching, pointing tasks, etc.); direction of view and locomotion control may be decomposed, etc. Nevertheless, during the experiments it was observed that most of the participants intuitively tend to control their locomotion also by moving their upper body, in spite of the inefficiency of such movements to significantly increase the inclination of the platform. However, this input (i.e., the orientation of the spine relatively to the hips) could be used to control navigation based on an holonomic locomotion model: as opposed to the non-holonomic one, lateral velocities are allowed in addition to tangential and angular ones by removing the constraint imposed by (5.2)—lateral velocities here would correspond to side steps.

5.3.3.4 Immersion and sensory feedback

The evaluation showed that JoyMan provides a satisfactory level of immersion in the virtual world. In addition, the Presence criterion was better ranked for this interface, compared to the joystick. As future work, we envisage to improve the level of immersion by adding other interaction techniques. For example, the visual perception of motion could be improved by adding camera motions like in [294]. Having an oscillating view point may reinforce the accuracy of the perception of the distance covered, by reproducing the natural oscillations of the head during locomotion. This effect could also reduce the sensation that locomotion is too slow, as reported by most participants in the subjective questionnaire.

5.3.4 Conclusion

In this section we presented JoyMan, a novel interface for the control of locomotion in virtual worlds meanwhile globally static in the real space. JoyMan is composed of a new peripheral device and a dedicated control law, which transforms the device state into a virtual locomotion velocity vector. Our main contributions are i) an exploration of the users' equilibrioceptive ability to control their virtual locomotion, and ii) the ability to maintain a high level of immersion compared to hand-held devices (e.g., joysticks).

The results of a preliminary evaluation are promising, since users enjoyed the navigation with the new interface. Sense of presence and realism of the virtual rotations were also assessed. This evaluation opens future directions of improvement and extension of JoyMan. Various VR, but also real world applications can be envisaged for its use: to cite some, videogames, rehabilitation, training, and virtual tours.

5.4 Shake-Your-Head: Walking in place using head movements

Shake-Your-Head is a new implementation of the WIP paradigm, designed to be compatible with desktop VR applications. Contrarily to traditional WIP techniques, Shake-Your-Head can be implemented using low cost devices.

5.4.1 Revisiting WIP for desktop VR

The whole pipeline of the WIP technique can be revisited, to match a larger set of configurations and apply it to the context of desktop VR. The approach is illustrated in Figure 5.5, highlighting the main differences of the new approach in front of existing WIP techniques.

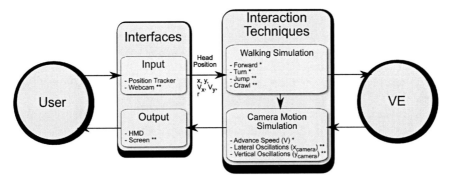

Fig. 5.5: Overview the proposed WIP approach (one star denotes improvement of existing components, two stars denote additional components).

With Shake-Your-Head, users can stand or sit such as in traditional video games or desktop VR configurations. They interact with the system by means of head movements. These movements can be captured using different tracking systems, particularly low-cost optical devices including standard webcams. On top of walking, the simulated locomotion affords also turning, jumping, and crawling. Finally, users can perceive such movements in the virtual world by means of integrated virtual camera motions on the three spatial axes, to further enhance the sensation of walking.

In the following we will describe the different components of our approach, namely: i) the 3D user input/output interface, ii) the interaction techniques for the computation of the virtual locomotion, and iii) the visual feedback based on camera motions.

5.4.1.1 Input/Output interface

Shake-Your-Head proposes new interface features extending the set of configurations where WIP can be applied, especially in Desktop VR. Specific elements in both the input and output interface are proposed to be incorporated for this purpose.

Input: Tracking based on head motions

The input interface is controlled by head motion. The current implementation consists of a webcam, allowing use in Desktop configurations without additional peripheral devices. The same interface can also be implemented using other traditional tracking systems.

Use of head movements

The concept is based on using head oscillations as cues for the identification of natural walking. During the walking act, in fact, the head oscillates along the lateral, vertical and forward axes [163]. The oscillations are strongly correlated to gait and footstep events. Moreover, these oscillations can also be measured while walking in place.

Head motions are traditionally recorded in existing WIP techniques thanks to the use of position trackers [273]. More in general, any tracking device can be used as long as its accuracy is within the range of 1 cm. Here, the use of a video camera is proposed to track the head position.

Extracted data

A three-DOF information set can be easily extracted by the camera images. Specifically:

- the lateral position x (and the computed speed V_x);
- the vertical position y (and the computed speed V_y);
- the rotation r of the head in the frontal plane.

These components are illustrated in Figure 5.6.

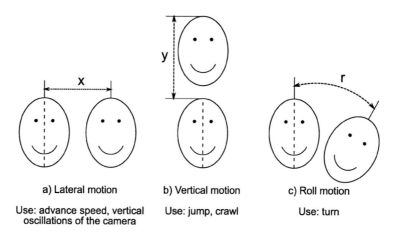

a) Lateral motion   b) Vertical motion   c) Roll motion

Use: advance speed, vertical   Use: jump, crawl   Use: turn
oscillations of the camera

Fig. 5.6: Extracted head motions: (a) lateral, (b) vertical, (c) roll.

Implementation

The use of the webcam to track the 3D position of the user head without markers requires the observation of real time constraints: in practice the algorithms must guarantee an acquisition rate at more than 25 frames per seconds. The Continuously Adaptive Mean Shift (Camshift) algorithm [44], implemented in the OpenCV library, is based on color detection and suited well for real-time tracking of colored objects and shapes. The algorithm recognizes a user's face as an ellipsoid. The head position (x, y) is identified as the center of the ellipse, and the angle r of orientation of the ellipse relative to the vertical axis is computed.

In general, x and y depend on the resolution of the specific webcam. Thus, a normalized position (x_n, y_n) ranging between -1 and 1 is determined on both axes. From the normalized values, the instantaneous speeds V_x and V_y can be computed; Kalman filters reduce the noise produced by the algorithm.

Output: Immersive and desktop visual displays

Both immersive and traditional screens can be employed as output interfaces. Concerning normal desktops, techniques can be used that tolerate a limited field of view. In WIP techniques, the field of view normally spans 360°, except for the case where redirected walking is used to simulate a 360° field of view in a four-wall CAVE [254].

Prospectively, Shake-Your-Head can be used with traditional output WIP interfaces such as HMD or CAVE systems. Here, it will be evaluated with either LCD laptop screens or video projections.

5.4.1.2 Walking simulation

Walking states

The main goal of Shake-Your-Head is to translate head motions into virtual motions inside the VE. In this environment the user should be able to perform various gestures while navigating. We implemented different locomotion states: walking, turning, jumping and crawling. To manage these states, we added a state automaton to the algorithm. The state transitions are governed by the head motions. The main inputs are the lateral velocity V_x and the vertical velocity V_y.

Forward state

Forward movements in the VE are controlled by lateral head oscillations. The idea is to enable walking with a variable locomotion speed, depending on the user's head motions. The speed V_a in fact is subjected to regular oscillations, that are proportional to the lateral head motion. One oscillation period in the head velocity corresponds to one step. Footstep events are associated to a null speed, corresponding to a change in sign of the lateral velocity. Thus, when the user's head reaches the amplitude peak of the oscillation then its velocity is null, as well as the locomotion speed.

Two thresholds, T_{min} and T_{max}, are introduced to obtain more realistic virtual walking. The former threshold flattens lateral head motions that are too small, hence avoiding forward locomotion under these situations. The latter prevents from reaching exceedingly high locomotion speed. The velocity V_a is computed in two steps while trimming such thresholds, according to the following equations:

$$V_{n_1} = \frac{\min\{|V_x|, T_{max}\}}{T_{max}}$$

$$V_{n_2} = \begin{cases} 0 & \text{if } V_{n_1} < T_{min} \\ V_{n_1} & \text{otherwise} \end{cases} \tag{5.8}$$

$$V_a = V_{n_2} \cdot V_{max}$$

Finally, the advance speed V of the camera inside the VE is adapted depending on the current locomotion state and is given by:

$$V = \begin{cases} V_a & \text{if state} = \text{walk} \\ 0.4 \cdot V_a & \text{if state} = \text{crawl} \\ V_{max} & \text{if state} = \text{jump} \end{cases} \tag{5.9}$$

Both thresholds were finally normalized to the following values: $T_{min} = 0.05$, and $T_{max} = 0.5$. We also set V_{max} to 3.5 m/s, corresponding to the highest possible speed.

Jump and crawl states

Compared to existing WIP techniques, two further locomotion states have been added to the navigation features in the VE: Jump and Crawl. These states are controlled by the vertical oscillations of the user's head. If the vertical velocity exceeds normalized thresholds T_{jump} in upward direction and T_{crawl} in downward direction, the user can jump and crawl respectively in the VE. In practice, this means that the user will need to slightly jump or bend forward if sitting, or jump or crouch down if standing. The user has to stand up to stop crawling.

When a jump is detected, the vertical position of the camera is set to follow a parabolic trajectory defined by:

$$y_{camera} = \frac{1}{2}gt^2 + Vt \tag{5.10}$$

with gravity acceleration g and time t. The jumping state is left automatically while landing, i.e. when the camera returns to its normal height H (known as the reference state when the algorithm starts). After preliminary testing, we set $T_{jump} = 0.3$ and $T_{crawl} = 0.4$. When the crawling state is active, the vertical position y_{camera} of the virtual camera is lowered by 1 m.

Turn state

In parallel to Forward, Jump and Crawl states, users have the possibility to modify their navigation direction. When performing a turn during normal walking, the human body leans to compensate the centrifugal force [66]. This phenomenon is often reproduced by videogame players, who tend to lean in the direction of the turn even if this gesture has no influence on their virtual trajectory. This feature is used in Shake-Your-Head to inform a control law based on the head orientation around the roll axis. To turn in the VE, users have to lean their head left or right. The rotation speed V_r of the virtual camera is given by:

$$V_r = \begin{cases} V_{r_{max}} & \text{if } r > r_{max} \\ -V_{r_{max}} & \text{if } r < -r_{max} \\ 0 & \text{otherwise} \end{cases} \tag{5.11}$$

where r_{max} is the minimum angle of head inclination enabling a rotation, and $V_{r_{max}}$ is the largest angular speed of rotation. In our experiment, we set $r_{max} = 15°$ and $V_{r_{max}} = 45°/s$.

5.4.1.3 Visual feedback based on camera motions

To further emphasize the sense of walking in the VE, the rendering of WIP was extended by using camera motions driven by the user's head oscillations. There are models in the literature, explaining how to enable camera oscillations along the three axes [163]. However, the oscillations are totally independent from the user control.

Here, a new model of camera motion control is adapted to the user's head movements. The camera oscillations along the three axes follow the user in real time to maintain the coherency of the system. In this way, visual feedback employing camera motions can be realized accounting for vertical, lateral and forward axes.

Forward oscillations

The forward speed V of the view point already oscillates. The camera motions are intrinsically linked to the advance velocity of the control law presented in 5.4.1.2. As a result, additional motion is not necessary along this axis and the camera advances at a speed that corresponds exactly to V.

Lateral oscillations

When walking, the users' heads oscillate left and right. Thus, as they move in front of the screen, their view point is modified by the head oscillations.

The lateral oscillations of the camera are computed as a function of the user's position. If d is the distance of the user from the screen and α and β are the aperture angles of the webcam, then the real position of the user in front of the screen depends on the normalized coordinates x_n and y_n. The real position of the user's head is given by the following coordinates:

$$\begin{cases} x_{real} = x_n \cdot d \cdot \tan(\alpha/2) \\ y_{real} = y_n \cdot d \cdot \tan(\beta/2) \end{cases} . \tag{5.12}$$

Finally, the virtual camera is moved along the lateral axis by a distance $x_{camera} = A_x \cdot x_{real}$. We set the scale factor A_x to 1 to match the user's head displacement, and thus generate the illusion that the screen is a window through which the user can observe directly the VE. However, other values can be used to scale the camera motions. The webcam parameters used during the experiment were set to $\alpha = 60°$ and $\beta = 45°$.

Vertical oscillations

The vertical oscillations of the camera can not be computed with the same algorithm as that for lateral oscillations. Users for instance can sit in front of a desktop, hence being unable to produce high vertical oscillations.

Shake-Your-Head generates pseudo-sinusoidal vertical camera oscillations based on the actual phase of the virtual gait cycle. Similarly to the advance speed control law, the vertical amplitude y_{camera} of the camera oscillations is given by:

$$y_{camera} = V_{n_2} \cdot y_{camera}^{max} \tag{5.13}$$

where y_{camera}^{max} is the amplitude of the vertical oscillations for speeds that are higher than or equal to the threshold T_{max}. For lower speeds, the amplitude of the oscillations is proportional to this maximum, thus increasing the perception of the variations in locomotion speed. The same factor between the camera motions and the locomotion speed forces the synchronization, resulting in a smoothed visual rendering. In our implementation, we set $y_{camera}^{max} = 15$ cm.

5.4.1.4 Discussion

To summarize, Shake-Your-Head is composed of i) an input interface based solely on the user's head movements, ii) a locomotion simulation in the VE composed of various possibilities such as jumping, crawling, turning, and iii) a visual feedback relying on oscillating camera motions. Head motions are tracked along three

DOF: lateral, vertical and roll axis (Figure 5.6). These different physical motions are transposed in virtual movements thanks to a locomotion automaton. We then included oscillating camera motions (5.9), (5.12) and (5.13) in the visual feedback, to enhance the walking sensation. The different control laws were parameterized after preliminary tests. Of course some parameters can be modified to amplify/decrease their effects during virtual locomotion. Besides this, other gesture-based control possibilities could also be envisaged, and added to the automaton such as running or walking backward.

5.4.2 Evaluation

Shake-Your-Head was evaluated by comparing it to a traditional desktop instead of alternative, more costly WIP techniques. In the latter setup, keyboard and joystick were provided to users, who were asked to navigate along a path whose complexity was given by a number of gates appearing on the way. Two paths were proposed, namely "normal" and "steeple".

Users tested Shake-Your-Head against the traditional input peripherals, either standing or sitting in front of the screen. In both positions, keyboard and joystick were employed to define control conditions in the experiment.

5.4.2.1 Task completion time

For each participant, task completion times were measured. A preliminary Principal Component Analysis revealed one outlier, who was excluded from the sample of the population. A specific analysis was carried out to investigate the existence of learning effects in the two conditions (Shake-Your-Head vs. joystick and keyboard). A linear model accounting for both conditions was fitted with the task completion times as functions of the trial number. The model revealed that the slope of the regression line was significantly lower than zero (p-value < 0.000001), reflecting a significant decrease in the completion time as the trial number increased. The same analysis, where the first trial was removed, conversely showed that the slope was not significantly different from zero. In consequence of this evidence, the first trial was removed from the analysis.

A two-way ANOVA was performed on the two conditions and for both positions. A post-hoc Tukey's analysis was then performed. Either path (normal and steeple) was dealt with separately by the ANOVA.

Concerning the normal path, the two-way ANOVA computed using the data on condition and position revealed a significant dependency between the position and the task completion time ($F(1,11) = 31.9981$, $p < 0.0001$) and between the condition and the task completion time ($F(1,11) = 6.0449$, $p = 0.0143$). The interaction between condition and position was also found to be a significant factor discriminating the completion time ($F(1,11) = 27.1921$, $p < 0.0001$). The post-hoc anal-

ysis showed that, while sitting with Shake-Your-Head, the task completion time
($M = 55.67$ s) was significantly lower than when sitting using keyboard and joystick ($M = 61.72$ s, adjusted p-value < 0.0001), when standing using keyboard and
joystick ($M = 62.06$ s, adjusted p-value < 0.0001), and finally when standing using
Shake-Your-Head ($M = 64.23$ s, adjusted p-value < 0.0001). The other combinations of interactions did not give significant adjusted p-values.

The results concerning the different conditions with the normal path are illustrated in Figure 5.7, ordered trial-by-trial respectively for both positions. The first
trial is included in both figures to show the learning effect.

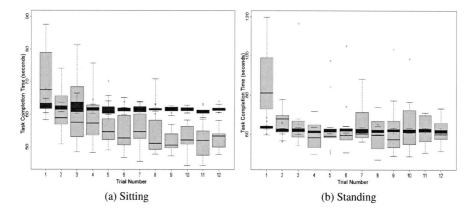

(a) Sitting   (b) Standing

Fig. 5.7: Task completion times on the normal path for Shake-Your-Head vs. keyboard and joystick obtained while (a) sitting, and (b) standing. Blue and red colors correspond to use of key-
board/joystick and Shake-Your-Head, respectively. Each box plot is delimited by the quartile (25%
and 75%) of the distribution of the condition over the individuals. The median is also represented
for each condition.

For the latter path, the pre-analysis again suggested the presence of a spurious individual and a learning effect affecting the first trial. The two-way ANOVA accounting for the conditions and positions revealed a significant dependency between the
position and the task completion time ($F(1, 11) = 8.5665$, $p < 0.005$), and between
the condition and the task completion time ($F(1, 11) = 11.8925$, $p < 0.001$). The interaction between condition and position was also considered as a significant factor
to discriminate task completion time ($F(1, 11) = 8.7647$, $p < 0.005$). A post-hoc
analysis showed that the task completion time when standing using Shake-Your-
Head ($M = 86.07$ s) was significantly higher than when sitting using keyboard and
joystick ($M = 74.78$ s, adjusted p-value < 0.0001), when standing using keyboard
and joystick ($M = 74.72$ s, adjusted p-values < 0.0001) and finally when sitting
using Shake-Your-Head ($M = 75.64$ s, adjusted p-values < 0.0001). The other combinations of interactions did not give significant adjusted p-values.

5.4.2.2 Accuracy

For each participant and for each trial, the percentage of errors for the two paths was measured separately.

- For the former (i.e. normal) path: 0.61% of error when sitting using Shake-Your-Head, 0% of error when sitting using keyboard and joystick, 1.39% of error when standing using Shake-Your-Head, 0.09% of error when standing using keyboard and joystick.
- For the latter (i.e. steeple) path: 20.49% of error when sitting using Shake-Your-Head, 7.81% of error when sitting using keyboard and joystick, 27.28% of error when standing using Shake-Your-Head, 13.19% of error when standing using keyboard and joystick. We found a significant effect between keyboard/joystick and Shake-Your-Head for the steeple path.

5.4.2.3 Subjective questionnaire

A questionnaire was proposed in which participants had to score the tasks under the different conditions and positions from 1 (low appreciation) to 7 (high appreciation), according to nine subjective criteria: (a) Fun, (b) Ease of use, (c) Intuition, (d) Accuracy, (e) Presence, (f) Walking realism, (g) Fatigue, (h) Cybersickness and (i) Global appreciation. Figure 5.8 shows the results. Scoring 7 for Fatigue and Cybersickness respectively meant that the technique did not induce fatigue and did not cause cybersickness at all.

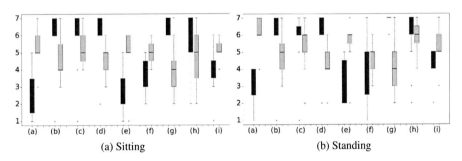

(a) Sitting   (b) Standing

Fig. 5.8: Subjective scores in (a) sitting and (b) standing positions. Blue and red light colors correspond to the keyboard/joystick and Shake-Your-Head condition, respectively. The subjective criteria are (a) Fun, (b) Ease of use, (c) Intuition, (d) Accuracy, (e) Presence, (f) Walking realism, (g) Fatigue, (h) Cybersickness, and (i) Global appreciation. Each boxplot is delimited by the quartile (25% and 75%) of the distribution of the condition over the individuals. The median is also represented for each condition.

Concerning the sitting position, no significant effect was found for Intuition ($p = 0.052$) and Cybersickness ($p = 0.12$). Concerning the standing position, no significant effect was found for Intuition ($p = 0.3$), Walking realism ($p = 0.19$) and Cybersickness ($p = 0.21$). We found a significant effect for all other criteria. In particular, Shake-Your-Head was better ranked for both configurations concerning Fun, Presence and Global Appreciation.

5.4.3 Discussion

The results suggest that Shake-Your-Head can allow efficient navigation compared to standard, well-known input devices such as keyboards and joystick. Some participants could go even faster with Shake-Your-Head, without significant loss of precision. Shake-Your-Head seemed also faster to learn. The technique was well appreciated and perceived as more immersive and more fun compared to traditional configurations.

The quick learning curve can be explained by the fact that the head movement control was intuitive and simple to learn [240]. After the learning phase, Shake-Your-Head in sitting position could allow faster locomotion compared to the traditional control. An explanation for the increased performance may be that with Shake-Your-Head, as well as with keyboard and joystick while standing, participants used to change direction without stopping their locomotion. Contrarily to this, with keyboard and joystick participants tended to walk and turn sequentially: this may globally increase the completion time. Another explanation could be that the locomotion speed with Shake-Your-Head was influenced by the speed of lateral movements. The sitting position allowed users to produce faster oscillations than while standing, thus speeding up locomotion. Interestingly, faster task executions were not detrimental to their precision.

The longer task completion time observed with Shake-Your-Head in the steeple paths may be due to unexpected virtual situations (i.e. jumping and crawling) inducing mistaken transitions in the automaton. Indeed, some participants apparently "prepared" the jump prior to doing it. The automaton could be easily evolved in the future, by including additional states accounting for both speed and user's position.

The results from the questionnaire are consistent with previous subjective evaluations, claiming that WIP techniques are more appreciated, more fun, and increasing feel of presence [304]. As expected, joystick and keyboard are found easier to use, more precise, and inducing less fatigue as they require less physical movement. Interestingly, cybersickness was not increased by Shake-Your-Head. This could be due to its use in the experiment jointly with a desktop (i.e. less immersive) configuration.

Finally, the realism of walking in the VE was significantly improved only in the sitting position. The perception of walking with Shake-Your-Head was quite complex: participants wrote "we have the impression to be a video game character", "the motions are exaggerated", or "we really have the sensation of walking, and not running". In the standing position, some participants found that physical motions were closer to "skiing" or "skating", as they noticed that they did not lift their feet from the ground but only oscillated with their body. To these people we could tell in the future that Shake-Your-Head still works well also if lifting their feet when walking in place, as the oscillations of the head are captured the same way either lifting one's feet or not.

Taken together, the results suggest that the Shake-Your-Head technique could be used in a wide range of 3D virtual navigation applications, with users in both sitting or standing position. It seems to be a low-cost and efficient paradigm that can afford several virtual locomotion situations. It could thus be used for training in VR with more physical engagement (military infantry, vocational procedures), or more realistic virtual visits such as for project review in architecture or urban planning.

5.4.4 Conclusion

We presented Shake-Your-Head, a technique designed to revisit the whole WIP paradigm to match a larger set of configurations, and apply it to the context of Desktop VR. Shake-Your-Head acquires solely the head movements of the user. It can be used in a desktop configuration with the possibility for the user to sit down and to navigate across the VE, through small screens and standard input devices for virtual tracking. Various locomotion situations have been implemented, such as turning, jumping and crawling. Additional visual feedback based on camera motion was also introduced to enhance the sense of walking.

An experiment was conducted to evaluate Shake-Your-Head compared to standard controls such as keyboard and joystick. In this experiment, participants had to walk through a series a gates. The evaluation was performed in both an immersive and desktop configuration. It was found that Shake-Your-Head only requires a short learning time to allow faster navigation in sitting position compared to keyboard and joystick. Moreover the technique was more appreciated, rated to be more fun, and providing more sense of presence compared to traditional techniques.

5.5 Magic Barrier Tape: Walking in large virtual environments with a small physical workspace

The Magic Barrier Tape aims at bringing a solution to immersive infinite walking in a restricted workspace, through the use of a natural and efficient metaphor.

5.5.1 The Magic Barrier Tape

Walking workspaces in VR systems are typically bounded by the tracking area, the display screens, or the walls of the immersive room. The Magic Barrier Tape has two basic objectives. The former is to display the limits of the workspace in a natural way, without interruptions of the immersion meanwhile avoiding collisions with physical objects outside the workspace boundaries or the tracking area. The latter is to provide integrated navigation furthermore affording any location in the virtual scene, also beyond the walking workspace.

To overcome the mismatch between the limited size of the workspace and the potentially infinite size of the virtual scene, the concept of hybrid *position/rate control* has been adapted from the realm of object manipulation inside an available workspace, where position controls fine object placement whereas rate control is used for coarse positioning at the boundaries [80]. This concept can be also found in common desktop applications and games, in which the mouse switches to rate control when the edge of the screen is reached: in a file manager when doing multiple selections, or in top-view strategy games such as Starcraft when panning on the map. In our context, the available workspace is the physical walking area. The workspace boundaries are represented by a virtual barrier approximately at the belt level, colored with slanted black and yellow stripes evoking the use of barrier tape, and its implicit message: "Do not cross".

The real workspace, delimited by the physical boundaries, hence is mapped to a virtual workspace inside the scene delimited by the virtual barrier tape. Inside this workspace, position control is set: the user can walk freely, and objects inside the virtual workspace can be reached and manipulated through physical walking and usual gestures. When reaching the boundaries of the workspace, the system switches to rate control: users can walk farther inside the scene by "pushing" on the virtual barrier tape, hence shifting the limits of the virtual workspace. Then, they can perform accurate positioning and manipulation tasks at the new virtual location.

The Magic Barrier Tape does not depend on a specific technology. The concept can be implemented under different VR systems. Any worn object or body part can interact with the virtual barrier tape depending on the application, and the rate control law can be fitted to specific behavioral needs. In the remaining of this section we detail the concept. Then, the Magic Barrier Tape is instantiated on a VR environment displayed through an HMD, in which one hand of the user is tracked across a cylindrical space over an area of about 7 m^2.

5.5.1.1 Display of the workspace limits

The workspace boundaries are displayed through three complementary visual components: the *main* virtual barrier tape, the *warning* virtual barrier tape, and their *grey shadow* on the floor.

The main virtual barrier tape is presented in the form of a band matching the shape of the workspace boundaries, e.g. a square for a CAVE or a circle for a cylindrical tracking region. The band is positioned at a safe distance ahead of them, high enough from the virtual floor so that the user does not need to look down to see the barrier tape, and low enough so that it does not occlude the user's forward vision. The boundaries of the workspace are therefore clearly and continuously visible. The tape is rendered as to appear semi-transparent, in ways to let the background be visible through the tape.

The warning virtual barrier tape appears when the user's body is close to the main tape, as a warning signal. This second tape has the same shape as the main one, and a red glow to capture the user's attention. For this reason, it is elevated at the eye level. This tape is fully transparent when the user is at a reasonably safe distance from the main tape, and becomes progressively opaque as the user gets closer, therefore making the warning signal also progressive, i.e. from dimmed to strong. The warning virtual barrier tape is complementary to the main tape, since it is triggered for safety measure, furthermore it gives an idea of when walking will be replaced by plodding.

Both tapes project a unique shadow on the floor, as if the tapes were illuminated from above. The shadow provides a visual cue about the limits of the workspace, visible when the user looks below. At least one of the three visual components is visible from almost any point of view; this is particularly helpful with an HMD setup, that usually displays a narrow field of view. Figure 5.9 shows the three components of the Magic Barrier Tape: the main barrier tape, the warning barrier tape, and their shadow.

Fig. 5.9: The three Magic Barrier Tape visual components showing the workspace boundaries: the main virtual barrier tape (middle), the warning tape (top) and their shadow (bottom).

In the proposed implementation, the main virtual barrier tape is 30 cm high and 30 cm far from the boundaries. It is shaped as a ring centered around the tracking area, with a 1.2 m radius. The ring is 1.3 m far from the virtual floor. The warning tape is activated when the user is 30 cm far from the main tape.

5.5.1.2 Navigation through rate control

The Magic Barrier Tape is based on position control inside the workspace, and rate control at the boundaries. This control is switched whenever the user's hand (or any other tracked body part) crosses the boundaries represented by the virtual barrier tape. The speed of the resulting translation in the virtual scene is a function of the hand penetration distance. When the user's hand returns inside the workspace, the user is switched back to position control.

Both the main and warning tape are stretched when a user's body part crosses the boundary. Their elastic behavior allows users to get a measure of how heavily they are plodding, and therefore to evaluate how fast they will move in the virtual scene. Some kind of visual feedback on the rate control is also important, so that the user can know where the neutral position is located [80].

The deformation D follows the shape of a centered Gaussian curve of equation:

$$D(p) = p \frac{1}{\sigma\sqrt{2\pi}} e^{-\frac{x^2}{2\sigma^2}}$$

where p is the penetration length (in meters), and σ is the std controlling the sharpness of the deformation. The virtual barrier tape rotates so that the center of the curve matches the penetration point P, the collision point between the hand and the virtual barrier tape. Therefore, as shown in Figure 5.10, the Gaussian deformation is centered around the penetration point, and its symmetry axis is given by the $\overrightarrow{OP}$ direction, where O is the center of the virtual barrier tape. Since the deformation

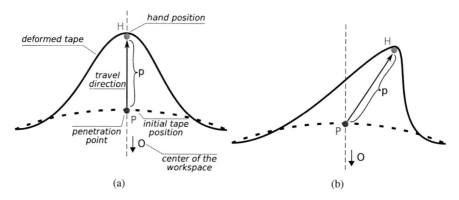

(a)   (b)

Fig. 5.10: The Gaussian deformation of the Magic Barrier Tape in top-view (a) and its shifted version (b) to follow the hand position.

follows the user's hand, the curve has to be shifted to take into account the lateral deviation of the hand position H with respect to the axis, as shown in Figure 5.10. The result models an elastic region around the penetration point, than can be deformed toward all directions. The deformation direction, $\overrightarrow{PH}$, corresponds to the shifting direction of the virtual workspace.

The velocity V gives the speed of this shift as a function of p:

$$V(p) = k * p^n$$

where k and n are constants. The use of a polynomial function results in low speed when the user is close to the boundaries, to cover short distances, and in high speeds to move fast, for targets that are far away. After preliminary tests, the current implementation employs the following parameters: $\sigma = 0.15$, $k = 1.4$ and $n = 3$. Such settings ensure a safe walking environment as well as a natural and efficient navigation.

5.5.1.3 An extension of resetting techniques for omni-directional walking

The evaluation of the Magic Barrier Tape was inspired by existing experimental methods furthermore suitable for testing this navigation technique. Among such methods, applicable in particular to real walking, only the *resetting techniques* developed by Williams et al. [327] account for collision free and infinite navigation scenarios. However, resetting techniques were originally designed for straight trajectories and 90° turns, conversely in most VR applications the user is allowed to explore the VE by taking arbitrary directions. Since the Magic Barrier Tape enables freedom of navigation, visual cues were added to extend such resetting techniques to omni-directional navigation.

Extended freeze-backup technique

In the original Freeze-backup technique, users reset their position by walking backward along a straight trajectory until reaching the resetting position. Since vision is inherently occluded during backward walking, resetting paths can only be straight. Otherwise, users could reach the workspace boundaries prematurely and find themselves "locked" in a too short resetting loopback.

In the extended Freeze-backup technique, users are taken back to the center of the real workspace. Before the reset, they can be positioned as well as oriented anywhere in the real workspace. Visual cues are thus added to guide the users to the resetting position, in two steps. First, their body is oriented: An horizontal segment is drawn on the screen, representing the user's orientation with respect to the resetting position. Users have to rotate until the segment becomes parallel to their body. Second, users have to walk to the resetting position by following the direction given by an arrow; this arrow signals also the distance from the target, by becoming smaller as the user gets closer to the resetting position. Through this mechanism, the user can reach the center of the real workspace from anywhere in the physical space.

Extended 2:1-turn technique

In the original 2:1-turn technique, a 180° physical rotation of the user is mapped into a 360° virtual turn in ways that users do not change their virtual direction. For this reason, walking paths must be straight with eventual turns avoiding "deadlock" problems as those mentioned above.

In the extended 2:1-turn technique, real turns must not necessarily amount to 180°. The resetting angle is in fact given by the (non oriented) angle between the physical and virtual orientation, also called resetting position vector. Conversely, the virtual angle constantly amounts to 360°. Before resetting, both turning directions are possible. The one forming the wider rotation angle is chosen, so that its mapping to a 360° virtual turn is smoother, and the consequent perceptual distortion is more tolerable. An arrow on the screen points the user to the more convenient turning direction.

5.5.2 Evaluation

The Magic Barrier Tape was tested against the two navigation techniques mentioned in the previous section [327] with extensions for omni-directional walking. Two experiments were conducted: a pointing task and a path following task.

5.5.2.1 Experiment #1: Pointing

The goal of experiment #1 was to compare the three techniques an a common pointing task, where the user had to move as rapidly as possible from a central (initial) to a new (target) location. It was assumed that the Magic Barrier Tape could afford faster navigation, since rate control allows to navigate at speeds that are greater than the average walking speed.

While comparing the analyses, a correction for experiment-wise error was done by using a Bonferroni-adjusted alpha level ($p = 0.05$ divided by the number of tests). In our comparison of the Magic Barrier Tape against the two other techniques, the alpha level was hence adjusted to $p = 0.025$.

Completion time

The completion times (in seconds) collected during 18 trials formed the data set for the statistical analysis. For each participant, mean and std were computed in each condition. A one-way within-subject ANOVA on the mean completion time revealed a significant effect of the Magic Barrier Tape ($F(2, 22) = 183.22, p < 0.001$). Subsequent t-tests revealed that the completion time in the Magic Barrier Tape (mean = 6.37 s, std = 1.30 s) was significantly shorter compared to the Freeze-backup technique (mean $= 21.49$ s, std $= 3.11$ s, $t(11) = -19.15$, $p < 0.001$). Similarly, the completion time in the Magic Barrier Tape technique was significantly shorter compared to the 2:1-turn technique (mean $= 14.54$ s, std $= 2.41$ s, $t(11) = -14.61$, $p < 0.001$).

Physical walking distance

An ANOVA on the mean physical walking distance (in meters) revealed a significant effect of the Magic Barrier Tape ($F(2,22) = 434.75$, $p < 0.001$). Subsequent t-tests revealed that the physical walking distance with the Magic Barrier Tape (mean $= 1.46$ m, std $= 0.16$ m) was significantly shorter compared to the Freeze-backup technique (mean $= 4.42$ m, std $= 0.30$ m, $t(11) = -30.13$, $p < 0.001$). Similarly, the physical walking distance with the Magic Barrier Tape technique was significantly shorter compared to the 2:1-turn technique (mean $= 3.37$ m, std $= 0.23$ m, $t(11) = -20.80$, $p < 0.001$).

5.5.2.2 Experiment #2: Path following

In the second experiment, the goal was to compare the three techniques on a path following task. The user had to walk as fast as possible meanwhile avoiding to go off a path, which was delimited by walls. As for the experiment #1, it was assumed that the Magic Barrier Tape could afford faster navigation, furthermore less precise path following during rate control [333].

While comparing the analyses, a correction for experiment-wise error was done by using Bonferroni-adjusted alpha level analogously to experiment #1. Also in this case, the alpha level was finally adjusted to $p = 0.025$.

Completion time

An ANOVA on the mean completion time (in seconds) revealed a significant effect of the Magic Barrier Tape ($F(2,22) = 84.01$, $p < 0.001$). Subsequent t-tests revealed that the completion time with the Magic Barrier Tape (mean $= 31.62$ s, std $= 9.71$ s) was significantly shorter compared to the Freeze-backup technique (mean $= 99.54$ s, std $= 21.63$ s, $t(11) = -12.06$, $p < 0.001$). Similarly, the completion time with the Magic Barrier Tape was significantly shorter compared to the 2:1-turn technique (mean $= 52.33$ s, std $= 6.59$ s, $t(11) = -6.48$, $p < 0.001$).

Path deviation

An ANOVA on the mean path deviation (in square meters) revealed a significant main effect of the Magic Barrier Tape ($F(2,22) = 4.77$, $p = 0.019$). Subsequent t-tests revealed that the path deviation in the Magic Barrier Tape (mean $= 3.46$ m^2, std = 1.76 m^2) was not significantly different compared to the path deviation in the Freeze-backup technique (mean = 2.45 m^2, std = 1.04 m^2, $t(11) = 1.72$, $p = 0.1143$). By contrast, the same analysis showed that the path deviation in the 2:1-turn technique (mean = 1.93 m^2, std = 0.54 m^2) was significantly lower than with the Magic Barrier Tape ($t(11) = 2.81$, $p = 0.017$).

Physical walking distance

An ANOVA on the mean physical walking distance (in meters) revealed a significant main effect of the Magic Barrier Tape ($F(2,22) = 379.81$, $p < 0.001$). Subsequent t-tests revealed that the physical walking distance with the Magic Barrier Tape (mean = 6.81 m, std = 1.33 m) was significantly shorter compared to the Freeze-backup technique (mean = 19.03 m, std = 1.25 m, $t(11) = -32.63$, $p < 0.001$). Similarly, the physical walking distance with the Magic Barrier Tape was significantly shorter compared to the 2:1-turn technique (mean = 13.61 m, std = 1.54 m, $t(11) = -13.17$, $p < 0.001$).

5.5.2.3 Subjective questionnaire

After both experiments, a preference questionnaire was proposed in which participants had to score (from 1 to 7) the three techniques according to six subjective criteria: Ease of use, Fatigue, Navigation speed, Navigation precision, General appreciation and Naturalness. Figure 5.11 shows the mean and std of the three techniques for each of the subjective criteria.

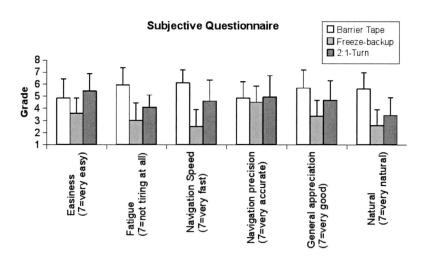

Fig. 5.11: Mean and std of subjective scores about the different criteria for the three techniques.

Wilcoxon-signed rank tests with Bonferroni correction showed significant differences: concerning Fatigue, between the Magic Barrier Tape and the Freeze-backup technique ($z = 2.69$, $p = 0.007$) as well as between the Magic Barrier Tape and the 2:1-turn technique ($z = 2.41$, $p = 0.016$); concerning Naturalness, between the Magic Barrier Tape and the Freeze-backup technique ($z = 2.77$, $p = 0.006$) as well as between the Magic Barrier Tape and the 2:1-turn technique ($z = 2.53$, $p = 0.011$); concerning Navigation speed, only between the Magic Barrier Tape and the Freeze-backup technique ($z = 2.82$, $p = 0.005$); and concerning the General appreciation, only between the Magic Barrier Tape and the Freeze-backup technique ($z = 2.65$, $p = 0.008$).

5.5.3 Discussion

Both experiments showed that the Magic Barrier Tape affords faster walking speeds compared to the other two techniques. In Experiment #1, the trials were completed more than three times as fast with the Magic Barrier Tape compared to the Freeze-backup technique, and more than two times as fast compared to the 2:1-turn technique. In Experiment #2, completion time using the Magic Barrier Tape was also roughly three and two times as fast, respectively. This result, which is mainly due to the fact that there is no waste of time when resetting the position with the Magic Barrier Tape, is consistent with the answers in the questionnaire regarding the navigation speed with the different techniques (Figure 5.11). Furthermore, the control law allows navigation speeds that are greater than the average walking speed. Completion times may be further reduced by tuning the control law for higher speeds, although in this case the interaction with the Magic Barrier Tape could become problematic, as posed by three users complaining that acceleration was sometimes hard to control.

The experiments also showed that users walked for a shorter distance when using the Magic Barrier Tape compared to the other techniques; this was expected, due to the use of rate control at the workspace boundaries. However, an interesting observation arises when considering that all trials were completed significantly faster using the Magic Barrier Tape: If we compute the ratio between physical walking distance and completion time on a subject-by-subject basis, we obtain similar values in Experiment #1 and in Experiment #2 respectively for the Magic Barrier Tape (mean = 0.24 with std = 0.04, and mean = 0.22 with std = 0.036), the Freeze-backup (mean = 0.21 with std = 0.04, and mean = 0.20 with std = 0.037) and the 2:1-turn (mean = 0.24 with std = 0.04, and mean = 0.26 with std = 0.052) technique. Hence, the amount of "productive walking" with the Magic Barrier Tape, corresponding to effective virtual forward locomotion with respect to time, is the same as with the other techniques. If we consider that walking speeds are all identical with the three techniques, then we can conclude that users spend roughly the same relative portion of the total time doing productive walking with the Magic Barrier Tape, compared to the other two techniques.

Experiment #2 showed that the Magic Barrier Tape was less accurate, with a larger path deviation compared to the 2:1-turn technique—roughly twice as large. Firm conclusions cannot be drawn about the comparison with the Freeze-backup technique, which did not not give significantly different results. Overall, these experimental conclusions were expected. By its nature and design, the use of the Magic Barrier Tape is meant for coarse positioning: users point to a target until getting it inside their workspace, and then they can hit it by fine navigation through physical walking. As explained by one subject, asked about the strategies he used during the experiment: "I sent the barrier tape as far as possible without going into the walls, in order to take advantage of the workspace". However, path deviation could be improved by allowing users to customize their control law, like when they choose the mouse speed in their desktop computer. Moreover, Zhai [333] observed that, with sufficient training, rate control and position control can achieve similar performances. Hence, further user training on the Magic Barrier Tape at rate control might reduce the path deviation.

Overall, users rated the Magic Barrier Tape better on all criteria of the questionnaire yielding significantly different comparisons. Six subjects complained about feeling cybersickness when using the 2:1-turn technique, furthermore two subjects said they were loosing balance. Many subjects found the Freeze-backup technique exhausting and frustrating. It is also important to note that two subjects had a very hard time in using the Magic Barrier Tape. They adopted an inadequate strategy, and complained about the control law. A longer training, or more guidance on the strategy to adopt would have probably helped these subjects perform better. Both of them rated the Magic Barrier Tape worse than the other techniques, in all criteria of the questionnaire.

In a nutshell, the Magic Barrier Tape affords to walk faster than the Freeze-backup and the 2:1-turn techniques, conversely it is less precise when used at rate control. The 2:1-turn technique is the most precise, but seems to cause cybersickness as well as loss of equilibrium. There is a general dissatisfaction with the Freeze-backup technique, mainly due to its physical exertion and slow speed, leading to a frustrating experience. People generally prefer the Magic Barrier Tape, and find it more natural and less tiring.

5.5.4 Perspectives

Potential improvements of the Magic Barrier Tape can be suggested, based on the experiments previously seen. Users can shift the virtual barrier using in principle any body part such as their shoulders, pelvis or feet, i.e., all such body parts we naturally use when our hands are busy.

In their "Bubble" technique [80], consisting of a hybrid position/rate control haptic interaction for devices suitable for limited workspaces, Dominjon et al. successfully employed haptic feedback to represent workspace boundaries possessing virtual elasticity. In the RubberEdge technique [54], Casiez et al. used the passive force feedback of an elastic ring put on top of a tracking surface, such as a touch-pad, to allow users to switch from position to rate control when reaching the elastic boundaries. Analogously, the Magic Barrier Tape could be augmented with haptic feedback anytime a user "pushes" against the tape: One possibility could be to use passive force through tangible objects, such as barriers with retractable belts as those found in front of ticket offices, airport check points, and in general wherever people must queue. These barriers should follow the workspace boundaries, and the virtual tape may be aligned to their position. Since retractable belts are responsive to the applied force, the haptic feedback consequence of the elastic deformation of the virtual barrier would be in principle perceived as natural.

Many users complained against the excessive translation speed when using the Magic Barrier Tape, that made the acceleration too hard to control. A solution to this problem may be to use a step-wise approach. The control law, according to the hand penetration, may deliver different speeds corresponding to a human walking, jogging, or running. Varying human velocities step-wise could be employed when moving along a direction orthogonal to the body orientation. The translation speed would therefore be more predictable, although always limited by the running speed.

Finally, although the barrier tape was set to be semi-transparent to reduce visual occlusion, background objects might become hardly visible if their dominant color was close to the tape color. A way to enhance visibility in such cases would be to use different tape textures complementing the dominant colors of the surrounding environment, also to emphasize the own appearance of the Magic Barrier Tape.

5.5.5 Conclusion

This section introduced the Magic Barrier Tape, a new interaction metaphor enabling users, confined to a restricted walking workspace, to navigate across an indefinitely large virtual scene. Using this metaphor and its implicit signalling, the walking workspace was circumscribed by a barrier tape surrounding the virtual scene. The Magic Barrier Tape uses a hybrid position/rate control mechanism: physical walking is afforded inside the workspace, while rate control navigation is employed beyond the boundaries when "pushing" against the virtual barrier. Moreover, it naturally informs users about the presence of boundaries in their workspace, hence providing a walking environment free of collision and tracking problems.

We conducted two experiments, to evaluate the Magic Barrier Tape against state-of-the-art techniques furthermore adapted to omni-directional navigation. In the former experiment participants had to hit a target, whereas in the latter experiment they had to navigate on a scene by following a given path. Results showed that the Magic Barrier Tape afforded faster locomotion compared to the other techniques. In particular, the latter experiment confirmed that navigation at rate control with the Magic Barrier Tape is not intended for accurate path following, but rather for coarse positioning preliminary to fine locomotion through physical walking. Overall the Magic

Barrier Tape was more appreciated, meanwhile affording more natural and relaxed task executions.

Future work will focus on the use of haptic feedback for a more compelling and immersive experience, as well as on trimming step-wise locomotion speed values producing a more predictable and realistic navigation at rate control.

5.6 Conclusion

Finite workspaces in which users physically walk must be often negotiated, when designing VR setups. Physical walking is furthermore constrained and bounded by the walls of the simulation enclosure and/or the range of the tracking system. These limitations pose serious challenges in the design of virtual navigation systems.

In this chapter we presented three novel navigation techniques which, in different ways, deal with interactive virtual navigation in large virtual environments under workspace restrictions. JoyMan is an input device that brings the joystick concept to human scale size; it is based on user's equilibrioception, i.e., users can lean with the body to control their virtual locomotion speed and own orientation. With Shake-Your-Head, users can "walk in place" to control an indefinitely long virtual walk; interestingly, Shake-Your-Head can make use of low-cost visual input devices to acquire head motions when walking in place. Finally, the Magic Barrier Tape implements a virtual barrier displaying the workspace limits in the virtual world; this concept enables safe and optimal physical walking inside the workspace. Users pushing on the virtual tape can switch to rate-control navigation, and reach locations otherwise outside the physically allowed walking area. The main features of the three techniques are summarized in Table 5.1.

	Display	Type of walk	Cost	Control law
JoyMan	HMD, CAVE	Static	Average	Rate control
SYH	Desktop screen	Walking-in-place	Low	Position control
MBT	HMD	Real walk	High	Hybrid rate/position control

Table 5.1: Main features of the three presented techniques.

JoyMan and the Magic Barrier Tape are more immersive than Shake-Your-Head. Among the three techniques, the Magic Barrier Tape provides the most realistic feedback. Conversely, JoyMan simulates less realistic virtual walking, but it can be used in CAVE environments and its cost is moderate. Finally, the low cost of Shake-Your-Head makes it interesting for applications such as video games and large-scale training systems.

Chapter 6
Pseudo-haptic walking

M. Marchal, G. Cirio, L. Bonnet, M. Emily, and A. Lécuyer

Abstract Pseudo-haptic feedback allows the simulation of haptic sensations in virtual environments using only the visual modality along with properties of human visuo-haptic perception. This technique has mainly been used for providing illusory kinesthetic feedback to the hands. In this chapter, we propose an extension of the same concept to the feet. We introduce novel interactive techniques to simulate the sensation of walking up and down in immersive virtual worlds based on visual feedback. Our method consists of modifying the relative position and motion of a virtual "subjective camera", while the user is physically walking in an immersive virtual environment. Changes of the virtual point of view are proposed in terms of variations of the following three parameters: height, motion speed, vertical orientation. As said, the effects of these variations were tested in an immersive virtual reality setup in which the user could walk. In parallel, a desktop configuration where the user was sitting and controlled traditional input devices was also tested, and compared to the former configuration. Experimental results showed that the proposed visual techniques efficiently simulate bumps and holes located on the ground. Furthermore, a prominent "orientation-height illusion" was found, since changes in vertical pitch evoke variations in perceived height even if this parameter is not changed. These effects could be applied in various virtual reality applications to provide sensations of walking on uneven grounds, such as in the virtual navigation of urban projects, for physical training, and in video gaming.

6.1 Introduction

Virtual Reality technologies immerse users inside a 3D synthetic world, simulated in real time by a computer. In such a virtual world, the user is given the possibility to manipulate virtual objects, and/or walk and explore virtual scenes.

Surprisingly, most current VR setups restrict users to walk on flat workspaces. Whilst this might seem appropriate in the inside of virtual buildings or across virtual streets, it becomes suddenly unrealistic for most outdoor walking experiences, such as when exploring a natural landscape. One main reason for this lack of irregular ground reproduction techniques lies in the current difficulty to simulate, in the physical workspace, uneven grounds by means of mechanically actuated interfaces. As of today, few achievements have been reported on the design of devices that can render uneven grounds such as locomotion interfaces [122, 123, 138]. For the moment these interfaces remain costly, cumbersome, and hence not widespread.

In video games, the user is generally sitting and acting through traditional manual devices. Mouse and keyboard are often used to control an avatar, and walk in the 3D virtual world in "first-person view". In this case, a technique which is commonly employed when navigating on uneven grounds consists of constraining the motion of the virtual camera to follow the floor profile. The camera stands always at the same height with respect to the ground level. This results in a continuous change in height of the view point, as if the user was "sliding" on the virtual ground.

In this chapter we study the use of these visual techniques to simulate uneven grounds, and to provide the sensation of walking up and down in an immersive VE while walking on a flat, real floor. The proposed techniques make use only of visual feedback, and consist of modifying the camera motion as a function of the virtual ground profile. Three effects are proposed: (1) a modification of the height, (2) a modification of the motion speed, and (3) a modification of the orientation. These effects are implemented and tested in two different configurations. The former consists of an immersive virtual reality setup in which the user is really walking, while wearing an Head Mounted Display (HMD) in charge of displaying the aforementioned visual effects. The latter consists of a desktop setup where a user controls her own walking by mouse and keyboard, such as in video games. We use both these setups to evaluate the influence of the different visual effects and their combinations. The Desktop configuration can be considered as a control condition, in comparison with the immersive visual scenario affording physical walking of the user.

The chapter begins with a description of related work in the field of simulation of walking in virtual environments. Then, we describe our visual effects and how they were implemented for the simulation of two simple profiles: a bump and a hole. Finally, we describe the results of an experiment conducted to evaluate the efficiency of the proposed techniques for simulating uneven terrains. The chapter ends with a description of potential perspectives and applications.

6.2 Related work

As of today, the simulation of the physical sensation of walking on uneven grounds has mainly been proposed through locomotion devices. When using these interfaces, users are engaged in a walking task while their motion is compensated by a backward movement of the mobile part of the device. Hence, the interface directly controls the position of the user in the virtual world. Most of the devices try to enable natural walking. There is a significant amount of locomotion devices specifically designed for VR systems and exploration of virtual worlds, however most of them can only reproduce flat surfaces without obstacles. In parallel, walking over uneven terrains and cluttered environments is an everyday experience (for instance occurring when one walks up and down the stairs), critical on some occasions such as when exploring outdoor environments. To date, only a few systems are capable of simulating human walking on non-flat ground.

The Sarcos Treadport [124], a treadmill with a mechanical tether attached to the back of the user, is an example of an attempt to provide a feeling of climbing slopes. Originally, the mechanical tether was used to compensate missing inertial forces and to simulate obstacles in the virtual path by applying forces on the torso. The concept was then extended so that the tether could also render the forces required to simulate a slope [122, 123]. A force on the opposite direction of motion was used when simulating upside walking, with a magnitude equal to the horizontal component of the force in the real world case, and, analogously, a force was applied on the direction of motion when going downhill. Simulation of side slopes was also possible by applying lateral forces.

Leaving the kinesthetic simulation and entering the haptic realm, the ATLAS [206] treadmill, mounted on an actuated spherical joint, was able to provide slopes by allowing the pitching and rolling of the platform. With a different approach, the Ground Surface Simulator (GSS) [207] was able to simulate uneven terrain through a linear treadmill with a deformable belt. Six long platforms could locally raise the belt, allowing the display of small bumps up to a slope of 5 degrees. The Sarcos Biport and the GaitMaster [138], both made of foot motion platforms, could simulate uneven but not inclined floors.

While these devices reproduce irregular grounds to some extent, they all suffer from common limitations that restrict their use widespread, mainly due to their size and weight, cost, lack of accuracy and versatility. Therefore, they have not yet been widely adopted outside the laboratory. Other smaller, less complex and more affordable interfaces exist that compensate locomotion. Foot-wearable devices like the Powered-Shoes [139] and, more recently, the Gait Enhancing Mobile Shoe [73], compensate one's motion without involving the use of a bulky structure. However, they cannot render slopes or any kind of uneven terrain.

In order to simulate haptic sensations without mechanical feedback, other solutions have thus been proposed such as sensory substitution and pseudo-haptic feedback [160]. The existence of pseudo-haptic effects has been discovered in experiments where the motion speed of a mouse cursor was modified according to the "height" of a texture [161]: As the mouse cursor slided over an image representing the top view of a 2D profile, an acceleration (deceleration) of the cursor indicated a negative (positive) slope of this profile. Experiments showed that participants could successfully identify profiles such as bumps and holes by variations of the motion speed. Camera movements were also proposed as means to modify the movement sensations in virtual environments. For instance, Steinicke et al. [277] used redirected walking techniques to orient the user trajectory in the real environment. Lécuyer et al. [162] used camera oscillations to reinforce the sense of walking in VEs. However, these approaches were not targeted to reproduce walking on uneven grounds.

In some video games, the "first-person" speed of motion is varied proportionally to the slope of the ground surface, thus providing cues of inclination. This effect could be classified as pseudo-haptic. However, to the authors' best knowledge there has been no study about the influence of such visual effects on the user's perception of heights and slopes in virtual environments. Furthermore, these effects have never been implemented within an immersive VR setup affording physical walking.

6.3 Novel interactive techniques based on visual feedback

6.3.1 Concept

The objective of the proposed interactive technique is to reproduce sensations of walking on uneven ground without the use of haptic or locomotion interfaces. The main idea consists of modifying the motion of the *subjective camera*, corresponding to the visual point of view, while the user is walking in the VE. The concept is that of controlling the camera position and orientation, depending on the virtual ground profile that is displayed on either a desktop screen or an HMD. The camera motion, hence, is a function of the simulated height of the terrain on which the user is walking. The variations of the camera motion are used here to render the inclination of a slope.

Three different modifications of the camera motion have been studied: height, orientation and velocity. Their respective magnitude values are computed from the height of the 3D virtual environment. Thus, the techniques can be used to simulate any uneven terrain, assuming that we know its 3D profile. Depending on the user motion, the algorithm computes the camera motion iteratively. When the user is moving in the VE, a displacement is figured out and the camera motion is computed by this measurement. Then, the new position of the user is computed to figure out the next position and orientation of the camera. The techniques described here resemble recipes used in video games. However, unlike most gaming situations, the aim here is to superimpose them to the virtual scene when the user is physically walking.

6.3.2 Implementation

The proposed effects (Height, Orientation, Velocity) are displayed in Figure 6.1. The combination of the three effects was also implemented.

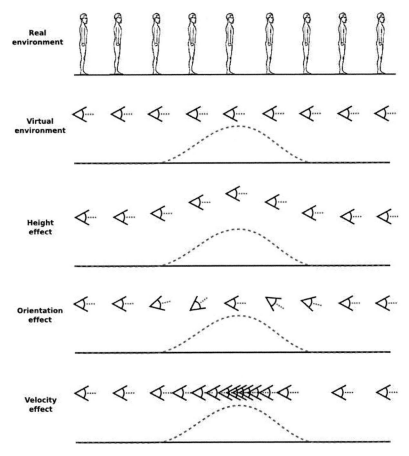

Fig. 6.1: The user is walking on a flat environment while encountering a bump. The camera view-point is modified in three different ways: Height (the camera slides with constant orientation and constant distance from the virtual ground surface), Orientation (the camera is kept parallel to the surface profile), Velocity (the camera motion speed decreases as the user is climbing a virtual bump, and increases when the user is walking down the virtual bump).

Height variation

The Height effect consists of modifying the subjective camera position by a translation along the vertical axis. This effect allows the user motion to be parallel to the ground surface during navigation. This approach is already used in a lot of demos and video games. The height varies following the equation:

$$\text{Height}^t = \text{Height}^{t-1} + \Delta_{\text{Height}} \cdot R_{\text{Height}} \qquad (6.1)$$

where Height^t is the vertical position at discrete time t. R_{Height} is the ratio applied to the difference of height Δ_{Height} between times $t-1$ and t. We introduced ratios in order to weigh the difference between subsequent heights. In our experiment, we chose $R_{Height} = 0.5$.

Orientation variation

The Orientation effect consists of applying a variation in the vertical orientation of the subjective camera, in order to create pitch changes when descending or ascending. This effect is supposed to mimic postural changes when walking on uneven grounds: the subject compensates the change in orientation and for this reason adopts a different posture. The camera angle at time t, $Angle^t$, is proportional to the tangential angle α^t at time t of the Gaussian curve:

$$Angle^t = \alpha^t \cdot R_{Orientation}. \tag{6.2}$$

Again, $R_{Orientation}$ is a ratio weighing the angle. In our experiment, we chose $R_{Orientation} = 0.5$.

Velocity variation

The velocity effect accounts for variations of the camera motion speed. In a real environment, a subject is generally moving slower on ascending slopes, and faster on descending slopes. We transpose this fact in our experiment by modifying the camera motion when the user is walking in a VE. Thus, the camera velocity is decreased when the user is walking up and increased when the user is walking down. This effect resembles the pseudo-haptic textures effect [161], here adapted to the simulation of first-person walking on uneven grounds. A different algorithm was used for the ascending and descending case. The algorithms compute at discrete time t the ratio $R_{Velocity}^t$ between the user's real velocity and the virtual camera speed. This speed is then modified by means of the equation

$$Velocity^t = Velocity^{t-1} \cdot R_{Velocity}^t \tag{6.3}$$

where:

- in the ascending case

$$R_{Velocity}^t = \exp(-R_{AscendingV} \cdot \alpha^t) \tag{6.4}$$

  with $R_{AscendingV} = 0.1$ in our experiments;
- in the descending case, the ratio $R_{Velocity}^t$ is designed to boost the effect immediately after the end of a bump or the beginning of the hole. At time t, this ratio is updated according to:

$$R_{Velocity}^t = R_{Velocity}^{t-1} + \Delta_{Height} \cdot R_{DescendingV} \tag{6.5}$$

  with $R_{DescendingV} = 2.0$ in our experiments.

When the subject reaches the end of the descending profile, her speed is at maximum. If the subject is walking across the hole, then she starts to go up and her speed value will be given by (6.4). If the subject has reached the end of a bump, $R^t_{Velocity}$ will start decreasing at 0.1 units per second until another bump/hole is reached or the ratio is set back to the initial value.

6.4 Evaluation of visual simulations of bumps and holes

The proposed visual techniques have been used to simulate bumps and holes having a Gaussian profile. This profile results in a distribution of height values along a line perpendicular to the walking path.

The investigation on the perception of 3D holes and bumps while walking in a VE was performed using an experimental protocol, consisting of the comparison of the different effects. The experiments were conducted using 3D virtual environments displayed either on a HMD or on a screen.

6.4.1 Virtual environment

The virtual environment consists of a simple corridor (height = 3.0 m, length = 19.0 m, width = 2.0 m). There is a region in the center of the corridor, whose height can be modified during the experiments: the user can walk either on a bump, a hole or a flat area. This region is highlighted by projecting a transparent solid on the ground having a height equal to 0.5 m and a surface equal to 3×2 m^2, as illustrated in Figure 6.2. The variable height of the ground is not visible, in order to exclude visual cues from the scene.

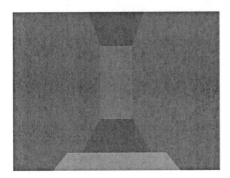 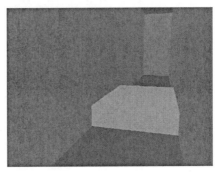

Fig. 6.2: (Left) Description of the virtual corridor; (Right) a transparent blue solid is placed in the bump region, to signal the area where changes in height occur.

6.4.2 Population and visual conditions

6.4.2.1 Group 1: Immersive VR configuration with HMD

Twelve participants (4 females and 8 males) aged 21 to 59 (mean = 28.7, std = 11.0) formed Group 1. One of them was left-handed, and none of them had known perception disorders. They were all naïve to the purpose of the experiment.

For this group, the experiments were conducted in an immersive room large enough to afford forward walking by about 6 m. An eMagin Z800 HMD was used to display stereoscopic images at 60 Hz. An opaque shield was mounted on top of the HMD, to prevent subjects from seeing the real surroundings. Users could move freely during the task, as illustrated in Figure 6.3. Their heads were tracked along all the task by an ART ARTtrack2 infrared tracking system, employing 9 cameras. The tracking region was 2.5 m high, 6 m long, and 3 m wide.

Fig. 6.3: Configuration of the immersive setup for the experiment using the HMD. In this picture, the scene displayed by the HMD is also projected on the screen to illustrate what the user is seeing during the experiment.

6.4.2.2 Group 2: Desktop configuration with monitor screen

Twelve participants (12 males) aged 21 to 59 (mean = 27.8, std = 6.1) formed Group 2. None of them participated also to Group 1. One of them was left-handed, and nobody had known perception disorders. They were all naïve to the purpose of the experiment.

This group joined an experiment made with a PC, whose keyboard was used to give the answers. There was no stereoscopic effect. Group 2 provided a control sample, useful to compare the immersive visual configuration affording physical walking against the desktop case.

6.4.3 Experimental plan

The displayed fields of view were identical for both the Desktop and HMD configurations. The goal was to evaluate and compare the effects of Height, Orientation and Velocity during the simulation of bumps and holes located on the virtual ground surface. We also evaluated a fourth effect, consisting of a combination of the three effects.

The experimental dimensions are the following:

- three *profiles*: Bump, Hole and Plane (i.e., flat surface);
- two *walking movements*: Forward and Backward;
- four visual *effects*: Height (H), Orientation (0), Velocity (V) and a combination of the three effects, namely HOV.

The experimental plan was made of the combinations [profiles x movements] across nine trials, for each effect, resulting in 54 trials per effect. Every subject alternated Forward and Backward movements, within a random sequence of [Bump, Hole, Plane] x [Forward, Backward] = 6 combinations. We performed a between-subject design. The 4 series (one for each effect) were presented using a Latin square and a defined sequence [H-O-V-HOV], counterbalanced with 4 sub-groups. The 12 participants of each group (Group 1 with HMD and Group 2 with PC) were thus equally divided into 4 sub-groups of 3 subjects each.

The motivation for testing backward movements is due to the fact that gait postures of human bodies are generally different when moving forward or backward on a slope. Thus, the hypothesis was that our visual effects could lead to different physical sensations in the case of backward movements .

6.4.4 Procedure

The experiment consisted of 216 trials per participant (54 per effect). Every subject had to go forward and then backward in the virtual corridor. At the end of each movement (either forward or backward), a black screen appeared on either the HMD or the screen, and the participant gave her answer concerning the shape identified (hole, bump, or plane).

6.4.5 Results

For each participant, the percentage of correct answers was estimated in the different experimental conditions. An ANOVA on the 4 different effects was conducted
on the percentage of correct answers. A post-hoc analysis using Tukey's procedure
was then performed. ANOVA were achieved separately for the two configurations
(HMD and Desktop) and by differentiating Forward and Backward movements. The
results concerning the different effects are represented in Figure 6.4 for the HMD
and the Desktop configuration. Results concerning Forward and Backward move-

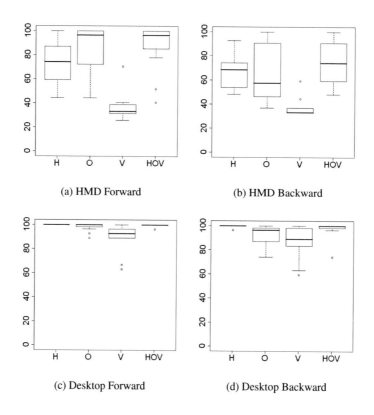

Fig. 6.4: Percentage of correct answers for HMD (above) or Desktop (below) configuration, as well
as for Forward (left) and Backward (right) movements. The effects H, O, V, and HOV, are represented on each box plot. Every box is delimited by the quartile (25% quantile and 75% quantile)
of the distribution of the effect over the individuals. The median is also represented for each effect.

ments are distinguished for each group, as they gave different values. The order in
the sequence had no significant effect on the results.

In the following paragraph, the results obtained for the four different resulting configurations are presented.

The ANOVA accounting for the four different effects revealed a significant dependency between the effect and the probability of giving a correct answer for all the configurations.

For Forward movements performed with the HMD configuration, the ANOVA performed between the four different effects revealed significant results for the effect $(F(3,11) = 19.447, p < 0.0001)$. Post-hoc analysis showed that the percentage of correct responses in the Height condition $(M = 73\%)$ was significantly higher than in the Velocity condition $(M = 37\%)$, adjusted p-value < 0.001; the percentage of correct responses in the Orientation condition $(M = 85\%)$ was significantly higher than in the Velocity condition, adjusted p-value < 0.001; and the percentage of correct responses in the HOV condition $(M = 87\%)$ was significantly higher than in the Velocity condition, adjusted p-value < 0.001. The other pairs of effects did not give any significant adjusted p-values.

For Backward movements performed with the HMD configuration, the ANOVA performed between the four different effects revealed significant results for the effect $(F(3,11) = 11.646, p < 0.0001)$. Post-hoc analysis revealed also significant differences: the percentage of correct responses in the Height condition $(M = 67\%)$ was significantly higher than in the Velocity condition $(M = 37\%)$, adjusted p-value $= 0.0004$; the percentage of correct responses in the Orientation condition $(M = 65\%)$ was significantly higher than in the Velocity condition, adjusted p-value $= 0.0010$; and the percentage of correct responses in the HOV condition $(M = 75\%)$ was significantly higher than in the Velocity condition, adjusted p-value < 0.0001. The other pairs of effects did not give any significant adjusted p-values.

For Forward movements performed with the Desktop configuration, the ANOVA performed between the four different effects revealed significant results for the effect $(F(3,11) = 7.77, p = 0.0003)$. Post-hoc analysis showed that the percentages of correct responses are significantly higher in Height condition $(M = 100\%)$, HOV condition $(M = 99\%)$, Orientation condition $(M = 98\%)$ than in the Velocity condition $(M = 89\%)$. Adjusted p-values are respectively 0.0008 (Height vs. Velocity), 0.0012 (HOV vs. Velocity) and 0.0072 (Orientation vs. Velocity). The other pairs of effects did not give any significant adjusted p-values.

For Backward movements performed with the Desktop configuration, the ANOVA revealed significant results for the effect $(F(3,11) = 3.8508, p = 0.016)$. Post-hoc analysis revealed that the percentage of correct responses are significantly higher in the Height condition $(M = 99\%)$ than in the Velocity condition $(M = 89\%)$ with an adjusted p-value equals to 0.015. The other pairs of effects did not give any significant adjusted p-values.

At first glance, regarding the percentage of correct answers for the different effects, it seems that the sensation of bumps and holes was perceived among the participants. HMD and Desktop configurations gave relatively different results. The Velocity effect with Desktop configuration gives namely higher percentages of correct responses compared to HMD configuration. Forward and Backward movements were distinguished for both configurations. Experiments conducted with a HMD and

Backward movement globally obtained lower results compared to the experiments conducted with the same experimental configuration but with Forward movements. We can particularly notice the lower results for the two effects containing the Orientation effect (O and HOV) for Backward movements.

For the HMD group, we can also notice the presence of two individuals (represented by individual dots on Figure 6.4 (a) and (b). These two individuals obtained lower percentages of correct answers for the HOV effect and higher percentages for the Velocity effect compared to the rest of the population, and should be considered as outliers.

An analysis concerning the percentage of correct answers for the different shapes identified (i.e. Hole, Plane and Bump) was conducted as well. The results are reported in Figure 6.5 for HMD and Desktop configurations, and detailed for Forward and Backward movements. Experiments conducted with the HMD contain more incorrect answers: the higher number of wrong answers for each effect occur with Plane shape for Height effect, as well as with Bump/Hole shape for Orientation and HOV effects. Thus, the Orientation effect seems to have an influence on the shape perception. For Velocity effect with HMD configuration, almost all answers were incorrect: Plane shape solution is almost always chosen, meaning that holes and bumps are almost never detected. On the opposite side, shapes with Velocity effect on Desktop configuration are well recognized. Thus, Velocity effect in an immersive situation leads to significantly different results, as observed also in Figure 6.4. Concerning backward movements with HMD configuration, we can notice that the percentage of incorrect answers is higher than for forward movements, for Height, Orientation and HOV effects.

6.4.6 Subjective questionnaire

After both experiments, a preference questionnaire was proposed in which participants had to rate from 1 (low appreciation) to 7 (high appreciation) the four different effects (H, O, V, HOV) according to 4 subjective criteria: ease of judgment, realism, cybersickness and global appreciation. Figures 6.6 and 6.7 show the results concerning the score obtained by the four different effects for each of the subjective criteria, for HMD and Desktop configurations.

As the number of modalities is high (7 levels), we assumed data to be normally distributed, hence an ANOVA on the four different effects was conducted on the score given to each criterion. ANOVA were performed separately for the two experimental configurations and were followed by a post-hoc analysis using Tukey's procedure. The ANOVA accounting for the four different effects revealed no significant dependency between the effect and its score for realism ($F(3,11) = 1.30$, $p = 0.28$) and cybersickness ($F(3,11) = 0.17$, $p = 0.91$) in the experiment made with the HMD.

Concerning global appreciation, the ANOVA performed between the four different effects revealed significant results for both configurations ($F(3,11) = 13.27$,

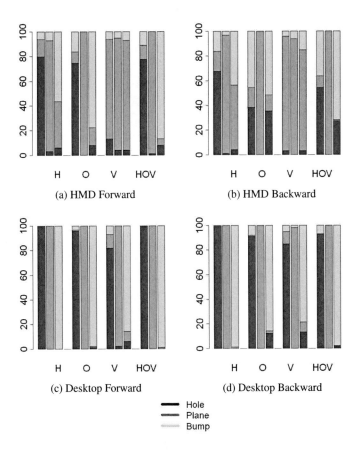

Fig. 6.5: Percentage of correct answers. Results are given for the 4 effects and the 3 different shapes (Hole, Plane and Bump in this order). The percentage of answers is decomposed for each shape, displaying additionally the incorrect shapes identified for each shape. Forward and Backward movements are distinguished.

$p < 0.0001$ for Desktop, $F(3,11) = 6.9$, $p < 0.0001$ for HMD). The HOV effect obtains the best global appreciation for HMD experiments, followed by Orientation and Height effects. Post-hoc analysis showed that Velocity is significantly less appreciated than HOV (adjusted p-value 0.0003) and Orientation (adjusted p-value 0.013). The other pairs of effects did not give any significant adjusted p-values, which argues in favor of a difference between the Velocity effect and the three other effects.

H obtains the best global appreciation in Desktop experiments, followed by HOV. Post-hoc analysis revealed that the Height effect is significantly more appreciated than Orientation (adjusted p-value 0.0008) and Velocity (adjusted p-value < 0.0001). The other pairs of effects did not give any significant adjusted p-values.

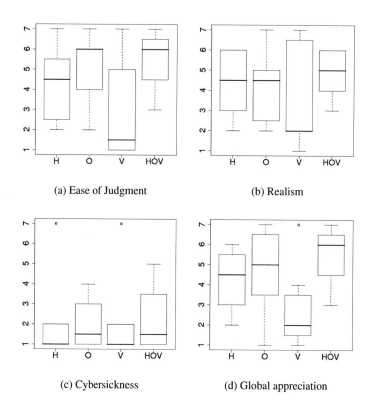

(a) Ease of Judgment   (b) Realism

(c) Cybersickness   (d) Global appreciation

Fig. 6.6: Subjective ratings for the four effects in HMD experiments: each box plot is delimited by the quartile (25% quantile and 75% quantile) of the distribution of the effect over the individuals. The median is also represented for each effect. The 4 different effects are represented on each picture: H, O, V, and HOV.

Indeed, Height was less appreciated with the HMD configuration. On the contrary, the other techniques were better accepted and fairly evaluated for the same configuration.

Concerning the ease of judgment, the ANOVA performed among the four different effects revealed significant results for both configurations ($F(3, 11) = 30.1$, $p < 0.0001$ for Desktop, $F(3, 11) = 4.53$, $p = 0.007$ for HMD). Post-hoc analysis for HMD configuration gave similar results as obtained for global appreciation. V is significantly less appreciated than HOV (adjusted p-value $= 0.0067$) and O (adjusted p-value $= 0.036$). The other pairs of effects did not give any significant adjusted p-values, which argues in favor of a difference between the Velocity effect and the three other effects. However, post-hoc analysis for Desktop configuration is slightly different than the one obtained for global appreciation. Concerning the ease of judgment, V is significantly less appreciated than the three other effects: H, O, and HOV (adjusted p-value < 0.0001 for all of them).

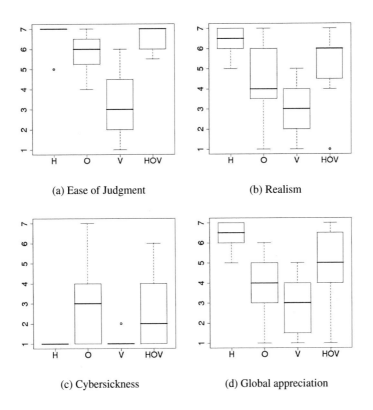

(a) Ease of Judgment   (b) Realism

(c) Cybersickness   (d) Global appreciation

Fig. 6.7: Results for subjective ratings about the different criteria for the four effects for Desktop experiments: each box plot is delimited by the quartile (25% quantile and 75% quantile) of the distribution of the effect over the individuals. The median is also represented for each effect. The 4 different effects are represented on each picture: H, O, V, and HOV.

Concerning cybersickness and realism, the ANOVA performed between the four different effects revealed significant results only for Desktop configurations ($F(3,11) = 6.56$, $p < 0.0001$ for cybersickness, $F(3,11) = 14.6$, $p < 0.0001$ for realism). Post-hoc analysis for realism revealed that H is more realistic than O (adjusted p-value = 0.0033) and V (adjusted p-value < 0.0001). Furthermore, HOV is more realistic than V (adjusted p-value = 0.0003). The other pairs of effects did not give any significant adjusted p-values. Post-hoc analysis for cybersickness showed significant adjusted p-values between O and H (adjusted p-value = 0.0025) and between O and V (adjusted p-value=0.0058). These results argue in favor of the exaggerated perceptions of the Orientation in Desktop configurations due to parameter values (i.e., the different ratios explained in Section 6.3), which seem to play a key role in the subjective appreciation of the participants.

Participants were also asked to evaluate the height of Bumps as well as depth of Holes during the experiment. Mean and std of their answers are given in Table 6.1

for both HMD and Desktop configurations. The real height/depth of the Bump/Hole

HMD				
	H	**O**	**V**	**HOV**
Bump	0.32 (0.28)	0.79 (0.56)	0.06 (0.15)	0.68 (0.45)
Hole	-0.3 (0.32)	-0.77 (0.55)	-0.05 (0.12)	-0.59 (0.40)
Desktop				
	H	**O**	**V**	**HOV**
Bump	0.59 (0.52)	0.97 (0.54)	0.42 (0.46)	1.32 (1.12)
Hole	-0.58 (0.52)	-0.99 (0.53)	-0.42 (0.46)	-1.28 (1.10)

Table 6.1: Mean and std (in brackets) of the heights/depths (in meters) given by the participants to bumps/holes for HMD configuration (above) and for Desktop configuration (below). The four different effects (H, O, V, HOV) are distinguished.

was 1.0 meter, with a ratio coefficient equal to 0.5 for Height, Orientation and Combination effects.

The estimated values are globally lower for HMD configuration. Interestingly, participants gave the Orientation effect the highest height for HMD configuration. The Orientation effect is always evaluated with an over-estimation of the correct height, although there is no variation in the camera height. The height values are under-estimated for Height effect for HMD configuration but slightly over-estimated for Desktop configuration. For Desktop configuration, the highest height is given to the experiments conducted with the HOV effect. For HMD configuration, the Velocity effect leads to a height value near to zero, but it is not the case for the Desktop configuration where the height change perception is better, as already observed in the results in Figures 6.4 and 6.5.

6.5 General discussion

At first glance, results show that slope was identified for most effects in the immersive configuration. Its appreciation greatly varies with the experimental setup and, to a lesser extent, with the motion direction.

The Height and the Orientation effects yielded highly positive results in an immersive configuration. Users clearly felt a change in height, and could distinguish in most of the cases whether it reported for a bump or a hole. In the Forward case for the immersive setup, the Orientation effect shows better results than the Height effect. Although there was no change in height, users were able to perceive it more accurately than in trials where the height itself changed. The success and accuracy of the Orientation effect was confirmed by the subjective questionnaire, since users had no trouble in sketching the profile of the shapes they encountered during the experiments. When estimating the height of these shapes, results were not so far from real heights. The sum of the effects did not give better results than one effect taken alone for HMD configuration. However, the HOV effect was more appreciated in the subjective questionnaire for the immersive situation.

On the other hand, Height was less appreciated in the HMD configuration. A possible explanation might be that users, particularly game players and people familiar with VR, are used to experience camera height variations when navigating in virtual uneven terrains in desktop environments. The Height effect is used in every desktop simulation involving slopes and landscapes. Rarely or never they have been exposed to the other effects. Hence, they find the Height effect more natural and acceptable. However, these same users have obviously less experience of immersive simulations, and might not be used to the conditions of an immersive setup. Consequently, they are less trained for the Height effect under these conditions and did not perceive any "real" sensation of floor inclination. Hence, the other techniques were better accepted and fairly evaluated in the immersive configuration.

When immersed in a VE with an HMD setup, it appears that users do not well perceive changes in height when exposed to the Velocity effect. Hence, one could think that the direct transposition of the pseudo-haptic effect from the 2D to the 3D realm does not provide the expected visual cues in an immersive configuration. Thus, when using the (non immersive) Desktop setup, the same Velocity effect, although not as effective as the others, yields much better results compared to the immersive setup, reaching an almost perfect score. A possible explanation for this behavior might be related to the optical flow of the virtual scene: In the immersive setup, the walls of the corridor were situated at the sides of the user's field of view due to the use of an HMD, while in the Desktop setup the entire display was largely contained in the field of view. Hence, the optical flow visible on the walls had a greater effect in the Desktop setup. The Velocity effect might have obtained better results with a different virtual scene and/or different HMD optics.

As expected, backward and forward movements led to different results. Profile recognition was worse for backward movements in the immersive configuration. A reason for this difference might be due to the tuning of the different parameters, especially when the Orientation effect is used.

The parametric setting for the different effects seems to play a key role in the recognition of bumps and holes. Parameters were set based on the HMD configuration. This obviously led to sub-optimal settings for the Desktop configuration, and could explain some differences in the results and subjective questionnaires (e.g. for the Orientation effect, for which the parameter value was probably too high). Future choices of the parameter values could be based on more accurate models. Although a very simple and straightforward implementation of the orientation motion was enough to achieve a good performance with the Orientation effect, other models closer to real walking scenes might improve the results. The physically-based model of an avatar representing the user in the virtual world, coupled to the motion of the user in the real world, might provide changes in head, and hence camera orientation, that are closer to what the user might expect. The use of real data of head orientation of users walking up and down slopes could also be an alternative solution to set these parameters. Finally, the use of other modalities can also be envisaged [299].

In future work, it could also be interesting to vary the steepness of the bumps and holes. Finally, a higher degree of realism, as called by many users in their subjective questionnaires, could improve the efficiency of the Orientation effect.

6.6 Conclusion and perspectives

In this chapter we introduced novel interactive techniques to simulate the sensation of walking up and down in immersive virtual worlds based on visual feedback. Our method consists of modifying the motion of the virtual subjective camera based on the variations in height of the ground surface level. This method has been widely used for instance in desktop video games, but never explored for providing realistic cues of ground profile to users walking in an immersive virtual environment. Three motion effects were proposed: (1) a modification of the height, (2) a modification of the motion speed, and (3) a modification of the orientation. They were tested in an immersive virtual reality setup in which the user was really walking. A Desktop configuration where the user was sitting and controlled input devices was also used to compare the results.

Experiments were conducted to evaluate the influence of visual feedback on the perception of virtual bumps and holes located on the ground. They showed that changes in height and orientation of the camera are effective for immersive VR. Motion speed seems to be less appreciated. Interestingly, in the immersive configuration, the coherent combination of all visual effects together led to the best result (although it was not found to be significant), furthermore this combination was preferred by the participants compared to the use of single effects. Experiments suggest also a strong perception of height changes caused by the orientation effect, holding the constancy of the camera height. This evidence was confirmed by subjective questionnaires, in which participants estimated a higher amplitude for bumps and holes simulated with the orientation effect. This "orientation-height illusion" opens intriguing questions in terms of human perception, and challenges our models of interpretation.

Taken together, the results suggest that the proposed visual techniques could be applied in an immersive virtual environment to simulate the sensation of walking on uneven surfaces. The same techniques could be used in various VR applications, such as in the virtual navigation of urban projects, for physical training, and in video gaming.

Chapter 7
Auditory rendering and display of interactive floor cues

S. Serafin, F. Fontana, L. Turchet, and S. Papetti

Abstract A walking task engages, among other senses, that of hearing. Humans do not only perceive their own footstep sounds during locomotion; Walking also conveys auditory cues that aid its recognition by listeners who are not performing the task. As a result, footstep sounds contribute to keep walkers in the perception and action loop, meanwhile informing external listeners about several characteristics of their actions. After reviewing the literature dealing with the auditory aspects of walking, this chapter provides an overview of some current developments in the interactive synthesis and display of footstep sounds. Several novel user evaluations conducted on auditory display prototypes applying these techniques are presented, shedding light on the potential of these techniques to find application in multimodal scenarios involving foot-floor interactions.

7.1 Introduction

During locomotion, humans are often engaged in parallel tasks distracting them from a thorough visual exploration and fine analysis of the floor that must be traversed. This is the case, for example, for a person walking in the city while browsing a newspaper, or simply looking at his or her surroundings. Symmetrically, it is not uncommon to notice walking persons who cannot continuously focus on a concurrent visual task, but who, due to their ability to exploit non-visual sensory cues, are seen to be capable of distinguishing the path ahead.

Such situations reveal a tendency by humans to keep walking tasks at the periphery of attention, a decision that is at any moment challenged by potential pitfalls that may occur during locomotion [211]. From this perspective, walking represents an interesting application domain in which to study possibilities for empowering the exchange of non visual information between humans and the external world.

The sonic modality, in particular, is at the core of many interactions involving communication at the periphery of one's attention. Physical contact with grounds that, once pressed or scraped, produce sounds, turn out to be especially informative. They also provide perceptual cues to walking persons nearby and to listeners who are outside this loop, but in the proximity of the interaction touchpoint. Such cues can reveal surface materials [111], shoe type and walking style, and can enable listeners to make inferences about the gender or other physical, biomechanical, and affective characteristics of a person [170, 309, 46].

This chapter builds upon a general hypothesis that subsumes the above situations: By providing peripheral, yet highly informative, cues, interactive walking sounds support listeners in their everyday activities involving self-locomotion.

This hypothesis is, in principle, applicable to a broad set of contexts and, hence, its implications have potential to span diverse social groups. Specifically, it would be interesting to assess walking sounds as human factors, and consequently measure their informative value in critical areas such as labor, locomotion in hostile spaces, or as navigation aids for people with different abilities [165]. Once available, these measurements may be useful toward enabling the design of auditory scenes in which such factors are optimized depending on the specific context of application.

In this chapter, we suggest that effective sonic interactions in walking can be realized through proper *augmentations* of the perceived reality. This means that future interactive floor scenarios should increase, rather than substitute the natural possibilities of grounds to convey auditory cues at floor level. In our case, additional cues will be enabled by enriching otherwise neutral floors with "layers" that provide physically-consistent sounds recalling familiar, possibly everyday ground ecologies. Examples include the superposition of sounds capable of evoking virtual aggregate grounds such as gravel, snow or mud.

Fortunately, current sensing and actuation devices allow interaction designers to come up with interface concepts, whose realization can be carried out at affordable cost and in reasonable time. Concerning, in particular, the floor interface domain, a plethora of force sensors as well as sound reproduction devices exist for detecting, then instantaneously responding to human walking through synthetic sonic feedback generated in real time by a laptop, or even smaller computing machines. In parallel it must be remembered that technologies differ in costs and feature specifications, and not all such technologies promise to scale down in price and encumbrance in their future versions. In the case of audio reproduction, the hardest constraints are perhaps given by the power consumption and weight of the amplifiers and loudspeakers. On one hand, we can opt for small, low-power devices providing fair quality and little low frequency content (bass sound), as a consequence of their limited acoustic bandwidth. On the other hand we can choose large amplification and reproduction devices with a broadband response, but bulky, by no means wearable auditory interfaces.

Whether the desired soundscapes will be finally achieved through rich and immersive, hence necessarily complex and encumbering systems or, conversely, through (even not computer-enabled) simple and miniaturized interfaces finding room in a shoe sole, is a question that will require time and further work to be answered. In this perspective, the chapter represents just a starting point toward that answer.

7.1.1 Chapter organization

This chapter discusses aspects of the design and realization process of soundscapes centered around walking sounds. In particular:

- Synthesis techniques are reviewed that allow one to simulate walking events, especially for the creation of virtual sonic layers of aggregate material on top of acoustically neutral floors. Since they are based on the simulation of simplified physical event descriptions, such techniques can be straightforwardly employed to synthesize vibrotactile signals that naturally arise out as a result of the same descriptions.
- Taken alone, a footstep sound has little perceptual meaning unless a suitable auditory context is present around it. This scene includes design of soundscapes that simulate different indoor and outdoor spaces.
- The resulting sonic scenarios must be rendered by adopting proper combinations of interactive walking sounds and soundscape descriptions. Not only they must achieve a sufficient degree of realism when displayed using a conventional reproduction arrangement, such as for instance a headset or a couple of stereo loudspeakers: they should be also flexible enough to accommodate unusual displays. This flexibility becomes especially interesting in the case of walking sound augmentations, for which some non-conventional reproduction sets are presented and then discussed.
- A user-centered evaluation of such scenarios is important toward assessing the ability of the simulations to realistically recreate ground surfaces.

7.2 Walking sound synthesis

7.2.1 Background

The first systematic attempt to synthesize walking sounds has been proposed by Cook in 2002 [65]. In this pioneering work, he introduced elements of novelty that make his work stimulating and still largely state-of-the art. The most interesting aspect in this modeling approach was the emphasis on foot-floor interactivity: thanks to an detailed procedure, which included several analysis stages, the model stored essential features from signals which were recorded during foot interactions with diverse floors; then, a reproduction of the same features could be made online by informing a parametric synthesis filter with temporal series of force envelopes, corresponding to footstep sequences. This allowed straightforward connection of the resulting system architecture to floor interfaces like sensing mats, performing a physically-informed interactive synthesis of walking sounds.

The physically-informed approach was also exploited in other work on the synthesis of walking sounds. By making use of physically-based algorithms for the reproduction of microscopic impacts [257], as early in 2003 Fontana designed a stochastic controller that, once parameterized in parameters of force and resistance (respectively against and belonging to the floor), generated realistic sound simulations of footsteps over crumples and similar aggregate materials. Thanks to a higher-level control layer, such sounds were grouped into a footstep sequence once they were triggered by an expressive control model exposing affective parameters and musical performance rules, proposed by Bresin. The resulting real-time software architecture was a document ("patch") for the Puredata software environment, which synthesized the sound and allowed continuous control of both physical floor parameters and gestural intentions of users [93].

An attempt to integrate some biomechanical parameters of locomotion, particularly the GRF, in a real time footstep sound synthesizer was made by Farnell in 2007 [91]. The result was a patch for Puredata that was furthermore intended to demonstrate how to create an audio engine for computer games in which walking is interactively sonified.

7.2.2 Synthesizing walking

Acoustic and vibrational signatures of locomotion are the result of more elementary physical interactions, including impacts, friction, or fracture events, between objects with certain shape and surface material properties such hardness, texture etc. The decomposition of complex everyday sound phenomena in terms of more elementary ones has been an organizing idea in auditory display research during recent decades [107].

Specifically, a footstep sound can be considered the result of multiple micro-impacts between a shoe and a floor. Either they converge to form a unique percept consisting of a single impact, in the case of *solid* materials, or conversely they result in a more or less dispersed, however coherent burst of impulsive sounds in the case of *aggregate* materials. At simulation level, it is convenient to draw a main distinction between solid and aggregate ground surfaces, the latter being assumed to possess a granular structure, such as that of gravel.

An impact involves the interaction between an active *exciter*, i.e., the impactor, and a passive *resonator*. Sonic impacts between solid surfaces have been extensively investigated, and results are available which describe relationships between physical and perceptual parameters of the objects in contact [147, 305]. Such sounds are typically short in duration, with sharp temporal onsets and relatively rapid decay.

The most simple approach to synthesizing such sounds is based on a lumped source-filter model, in which a signal $s(t)$ modelling the excitation is passed through a linear filter with impulse response $h(t)$ modeling the resonator, and resulting in an output expressed by the linear convolution of these two signals: $y(t) = s(t) * h(t)$.

By borrowing terminology from the kinematics of human locomotion, the excitation force can be identified with the GRF. In our case, GRF signals acquired using microphones or force input devices have been used to control different sound synthesis algorithms, which reproduce solid and aggregate surfaces as listed in the following of this section.

7.2.3 Physics-based modeling

The simulation of the interaction between solid surfaces can be obtained by decomposing the physical phenomenon into its basic constituents, instead of linearizing it into a series of two or more filters. The physics-based modeling approach precisely allows to deal with different kinds of interactions, by preserving their invariant phenomenological properties through this decomposition. According to this approach, situations like a foot sliding across the floor, or conversely walking on it, can be rendered respectively by starting from a friction or impact excitation component, meanwhile preserving the invariant floor properties in the resonant component.

Impact and friction are two crucial categories that affect walking perception [107]. In the impact model, the excitation corresponding to each impact $s(t)$ is assumed to possess a short temporal extent and an unbiased frequency response. A widely adopted physically-based description of this phenomenon considers the force f between the two bodies to be a function of the compression x of the exciter and velocity of impact $\dot{x}$, depending on the parameters of elasticity of the materials, masses, and local geometry around the contact surface [21]:

$$f(x,\dot{x}) = \begin{cases} -kx^{\alpha} - \lambda x^{\alpha}\dot{x}, & x > 0 \\ 0, & x \leq 0 \end{cases} \qquad (7.1)$$

where k accounts for the material stiffness, λ represents the force dissipation due to internal friction during the impact, α depends on the local geometry around the contact surface. When $x \leq 0$ the two objects are not in contact.

Friction has been already implemented in sound synthesis as well, by means of a dynamic model in which the relationship between relative velocity v of the bodies in contact and friction force f is represented as a differential problem [23]. Assuming that friction results from a large number of microscopic elastic bonds, also called bristles [82], the v-to-f relationship is expressed as:

$$f(z,\dot{z},v,w) = \sigma_0 z + \sigma_1 \dot{z} + \sigma_2 v + \sigma_3 w \qquad (7.2)$$

where z is the average bristle deflection, the coefficient σ_0 is the bristle stiffness, σ_1 the bristle damping, and the term $\sigma_2 v$ accounts for linear viscous friction. The fourth component $\sigma_3 w$ relates to surface roughness, and is simulated as fractal noise.

7.2.3.1 Interaction with aggregate surfaces

The sonic properties of aggregate surfaces can be reproduced by dense temporal sequences of short impact sounds. In most cases sound designers avoid modeling these properties at a fine-grained level of detail, since a profound inspection of the microscopic phenomenon does not bring proportional advantages to the quality of the synthesis meanwhile increasing the computational burden to levels that often become intolerable[1].

The crumpling algorithm [93, 47] implements a higher-level control, that is put on top of the physically-based impact-resonator model described above. In other words, it organizes temporal sequences of physical impacts between microscopic objects, assumed to be solid. Such sequences give rise to crumpling events, each represented by a corresponding group of micro-impacts. At the same time, the energetic evolution of a sequence instantaneously informs the parameters that are responsible for the generation of micro-impacts.

The temporal distribution of micro-impacts is governed by a Poisson stochastic process, whose inter-arrival times are given by the Poisson distribution $P(\tau) = \lambda e^{\lambda \tau}$ in which λ controls the stochastic density of the micro-impacts. In parallel, the power of each micro-impact follows a stochastic law $P(\gamma) = E^{\gamma}$, controlled by the γ parameter, which is derived from the physics of crumpling [268]. As a result, i) the dissipation of energy occurring during an impact, and ii) the temporal distribution of adjacent impacts can be constantly controlled (in stochastic sense) by the energy left to the process.

The crumpling model is characterized by the *average interval* between micro-impacts and the *average power* of every event as characteristic parameters for the control of sound. Once such controls are instantaneously mapped onto the force signals coming from a foot interface, the model allows continuous control over the generation of crumpling events that can be associated to footstep sounds [47].

Further mapping can be designed in between the interface and the above controls for setting the invariant features of an aggregate ground material, like its *resistance* or *compliance* parameters, having consequences in the perceived granularity of the ground. Together with proper settings of the modal resonator parameters defining the "color" of each micro-impact, these macroscopic controls set the acoustic signature of an aggregate material. Figure 7.1 illustrates the continuous crumpling algorithm.

7.2.3.2 Ground surfaces as resonant objects

In many cases of interest, a ground surface can be modeled as a linear resonator as opposed to the exciter. The ground properties determine the resonator parameters. Solid and homogeneous floors exhibit a narrow-band (hence longer and possessing definite color) sonic signature, conversely aggregate floors can be synthesized using bursts of short, wide-band (hence more noisy) resonant sounds simulating multiple collisions of the shoe against ensembles of small resonators.

---

[1] Note that this simplification cannot be used a general rule holding for all sounds resulting from multiple, small-scale processes. For instance, when substances with varying contact properties are involved such as liquids in motion, more sophisticated simulations must be realized which consider also the transitions across different macroscopic states [81].

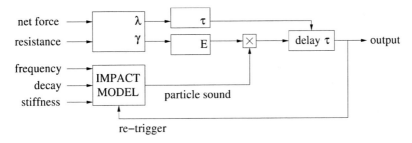

Fig. 7.1: Continuous crumpling algorithm.

Notably, the former sounds in practice are more difficult to be displayed. In fact, homogeneous floors are inherently large resonators capable of producing spatially diffused, possibly loud oscillations across a wide area (think of walking over a wooden floor or jumping on a metal grate). Conversely, aggregate materials generate localized sounds that normally can not propagate along the ground surface. The former, hence, need a powerful enough sound reproduction system: a constraint that rarely can be satisfied by a wearable interface.

As opposed to reproduction, the computational modeling of an even complex linear resonator is relatively straightforward. Modal synthesis [5] explains how to decompose a resonator that responds to an impulse with a signal $h(t)$ in terms of a number of resonant *modes*. The response, then, is modelled as a filter bank with impulse response $h(t) = \sum_i a_i e^{-b_i t} \sin(2\pi f_i t)$, where a_i represents the amplitude of the ith mode, b_i its decay rate, and f_i the associated modal frequency.

Each resonating filter is equivalent to a second-order damped linear oscillator:

$$\ddot{x}^{(r)}(t) + g^{(r)} \dot{x}^{(r)}(t) + \left[\omega^{(r)}\right]^2 x^{(r)}(t) = \frac{1}{m^{(r)}} f_{ext}(t), \tag{7.3}$$

where $x^{(r)}$ is the oscillator displacement and f_{ext} represents any external driving force, while the parameters $\omega^{(r)}$ and $g^{(r)}$ are the oscillator center frequency and damping coefficient, respectively. The parameter $1/m^{(r)}$ controls the "inertial" properties of the oscillator. Such a one-dimensional model provides a basic description of the resonator in terms of its pitch $\omega^{(r)}$ and quality factor $q^{(r)} = \omega^{(r)}/g^{(r)}$. The parameter $g^{(r)}$ relates to the decay properties of the impulse response of the system (7.3): specifically, the relation $t_e = 2/g^{(r)}$ holds, where t_e is the $1/e$ decay time of the impulse response.

7.2.4 Physics-based sound synthesis using the SDT

The Sound Design Toolkit[2] (SDT) [74] is a software product made up of a set of physically-consistent tools for designing, synthesizing and manipulating ecological sounds [106] in real time. The aim of the SDT is to provide efficient and effective instruments for interactive sonification and sonic interaction design.

The SDT consists of a collection of patches and *externals* for Puredata and Max/MSP.[3] The library is compatible with MacOSX, Windows, and Linux (Puredata only).

In the Puredata and Max/MSP terminology, an *external* is a dynamic library which provides some kind of signal processing. Depending on its communication interface (in the form of a set of *inlets* and *outlets*) an *external* can be linked to other *externals*, arithmetical operators, digital filters, sliders or other GUI elements that are natively provided by such environments. Together, all these elements find place inside patches allowing to define complete digital signal processing procedures.

In particular, each SDT's *external* represents a physically-based or -informed/-inspired algorithm for sound synthesis or control. The SDT patches make use of these *externals* to implement fully functional physically-consistent sound models. Moreover, they provide features for parametric control and routing of I/O signals.

The SDT has been used for synthesizing audio and vibrotactile feedback simulating different ground materials. Following is a brief description of the models, and how they have been used.

7.2.4.1 Realization of solid impacts

In the SDT implementation, a modal resonator can have an arbitrary number of resonant modes, each of which is represented by a linear 2nd-order oscillator in the form given by (7.3). Also, the resonating object can be endowed with a macro-dynamic behavior provided by an *inertial mode* added to the modal resonator structure. The inertial mode describes the macro-dynamics of a modal resonator as that of a pointwise mass, which is described by the Newton equation of motion:

$$\ddot{x}(t) = \frac{1}{m}f(t) \qquad (7.4)$$

where x is the *displacement* of the whole object, m is its *mass*, and f is the external *force* applied to the object. When present, the inertial mode is considered as the first mode of a modal resonator. It is clear that while an inertial mode is undamped, conversely resonant modes are damped. Having described its single components, it is now possible to describe the structure of a modal resonator having N modes of index $l = 1 \ldots N$ by means of the following linear system:

---

[2] http://www.soundobject.org/SDT/
[3] Which is in a sense the advanced, yet commercial, counterpart to Puredata.

$$\begin{bmatrix} \ddot{x}_1(t) \\ \vdots \\ \ddot{x}_N(t) \end{bmatrix} + G \begin{bmatrix} \dot{x}_1(t) \\ \vdots \\ \dot{x}_N(t) \end{bmatrix} + \Omega^2 \begin{bmatrix} x_1(t) \\ \vdots \\ x_N(t) \end{bmatrix} = \bar{m} f(t) \tag{7.5}$$

where G and Ω are diagonal matrices whose elements are, respectively: $g_{l=1...N}$ and $\omega_{l=1...N}$, and $\bar{m} = [1/m_{l=1...N}]^T$. In case the inertial mode was present, g_1 and ω_1 would be equal to 0, while x_1 would be the displacement of the entire object and m_1 its total mass.

The displacement x_j of a resonating object at a given point $j = 1...N$ can be calculated as:

$$x_j(t) = \sum_{l=1}^{N} q_{jl} \cdot x_l(t) \tag{7.6}$$

where the coefficients q_{jl} are the *output weights* for each mode $l = 1...N$ at the output point j. It is clear that, in case an input and an output point coincided (that is, $i = j$), their modal weights $1/m_{li}$ and output weights q_{jl} would also be the same.

The algorithms underlying solid surface sound models share a common structure which is shown in Figure 7.2: two objects interact through what is called an *inter-actor*, which models the actual contact interaction. The interactor contains most of

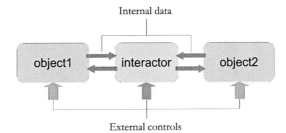

Fig. 7.2: The common structure underlying the SDT
algorithms simulating solid
surface contacts.

the art, that is necessary to couple two or more objects in the discrete-time domain.

In principle, in order to implement the algorithm represented in Figure 7.2 it is sufficient to couple two instances of (7.5) with (7.1) or (7.2). In practice, this elaboration must be performed while converting the same equations to the discrete-time domain. Thus, their coupling needs to solve issues about numerical stability and accuracy, as well as instantaneous propagation of the effects; for details, see [223].

In the current version of the SDT the bilinear transformation — an A-stable 2nd-order implicit method [244] — is used to translate the continuous-time equations above to the discrete-time domain. In order to discretize the system of 2nd-order differential equations of (7.5) it is useful first to rewrite a single mode (7.3) as an equivalent system of two 1st-order differential equations:

$$\begin{cases} \dot{v}(t) + gv(t) + \omega^2 x(t) = \frac{1}{m} f(t) \\ v(t) = \dot{x}(t) \end{cases} . \tag{7.7}$$

By applying the Laplace-transform to (7.7) we obtain two 1st-order equations in s. The next step is to apply the bilinear transformation, thus obtaining two equations in z, and finally apply the inverse Z-transform in order to obtain the following discrete-time system, expressed in *state-space* form:

$$\begin{bmatrix} x(n) \\ v(n) \end{bmatrix} = A \begin{bmatrix} x(n-1) \\ v(n-1) \end{bmatrix} + \begin{bmatrix} \frac{1}{4m\Delta} \\ \frac{F_s}{2m\Delta} \end{bmatrix} [f(n) + f(n-1)] \qquad (7.8a)$$

where the matrix A has the following expression:

$$A = \frac{1}{\Delta} \begin{bmatrix} \Delta - \omega^2/2 & F_s \\ -F_s\omega^2 & 2F_s^2 - \Delta \end{bmatrix} \qquad (7.8b)$$

with $\Delta = F_s^2 + gF_s/2 + \omega^2/4$.

As for the inertial mode, the discrete counterpart to (7.4) is easily obtained from (7.8a) and (7.8b) considering $g = 0$ and $\omega = 0$. It follows that the bilinear transformation enables to maintain a unified formulation for both the inertial and resonant modes.

Since the bilinear transformation is an implicit equation, the resulting discrete representation (7.8a) is also in implicit form: an instantaneous dependency between the state variables (displacement x and velocity v) and the input force f is present.

Finally, the discrete-time counterpart to the system made of (7.5) and (7.6), representing an entire modal object, can be written as:

$$\begin{cases} \begin{bmatrix} x_l(n) \\ v_l(n) \end{bmatrix} = A_l \begin{bmatrix} x_l(n-1) \\ v_l(n-1) \end{bmatrix} + \begin{bmatrix} \frac{1}{4m_l\Delta_l} \\ \frac{f_s}{2m_l\Delta_l} \end{bmatrix} [f(n) + f(n-1)] \\ \\ \begin{bmatrix} x_j(n) \\ v_j(n) \end{bmatrix} = \sum_{l=1}^{N} q_{jl} \begin{bmatrix} x_l(n) \\ v_l(n) \end{bmatrix} \end{cases} \qquad (7.9)$$

where $l = 1 \ldots N$ and $j = 1 \ldots N$ denote respectively the mode and output point considered. The matrix A_l is as in (7.8b), but now accounts for a different $\Delta_l = F_s^2 + g_lF_s/2 + \omega_l^2/4$ for each mode l.

Due to the implicit form of the bilinear transformation, the resulting discrete-time equations are implicit as well. Hence, an instantaneous relationship is present. For instance, in the case of the impact model, while the modal resonator of (7.9) needs f_{n+1} to compute $[x_{n+1} \; v_{n+1}]^T$, the impact force f_{n+1} also has an instantaneous dependence on x_{n+1} and v_{n+1} given by the discrete-time counterpart of (7.1). Such a *delay-free loop* is not directly computable and, because of the non-linear dependence $f(x,v)$, it needs some special handling in order to be solved. In particular, the *K-method* [42] together with *Newton's method* [244] are used.

The algorithm summarized in Figure 7.2 can now be seen in more detail: at each temporal step the resonators send their internal state (namely, displacement and velocity at the interaction point) to the interactor, which in turn, after solving the delay-free loop as explained above, can send the newly computed interaction forces back to the objects, thus putting them in condition to perform a computation for the next step. The non-linearities provide richness and improved dynamics to the resulting sounds, even when using low-order resonators.

The solid surface sound models from the SDT allow to set the number of modes of a modal resonator, and the control parameters allow manipulate their modal properties individually.

7.2.4.2 Application to footstep sounds

The SDT realization of solid impacts is a basic building block for synthesizing footstep sounds. A realization of the friction model (7.2) exist in the same library as well, whose discrete-time implementation is left out of this chapter for the peculiar numerical issues that it raises [23].

For the sake of footstep sound synthesis, SDT has been enriched with an alternative impact model implementation, called soft impact. This model allows to synthesize the sound of an impact on a soft surface, or a soft impact between two surfaces. Although avoiding an accurate simulation of the physics of contacts between spatially distributed objects, nevertheless the soft impact algorithm provides effective acoustic results by making use of a dense temporal sequence of tiny signals that excite the resonator described by (7.9). In more detail, no mutual interaction among an interactor and resonating objects is simulated. The interactor, i.e. the force f of (7.9), is instead substituted by a proper static force in the form of a noise burst that finally excites a modal resonator.

This algorithm, hence, realizes a simple feed-forward signal processing procedure. In spite of its simplicity, the idea behind can be qualitatively justified considering that smooth contacts can be reduced to dense temporal sequences of micro-impacts, in a sense modeling the surfaces of the interacting objects as multiple contact areas. Also, the use of specifically filtered noise signals can be motivated considering that such micro-impacts can exhibit a stochastic-like distribution. Besides the modal resonator parameters, the available controls include an ADSR (*attack time, decay time, sustain gain, sustain time, release time*) envelope shaper and the *cutoff frequency* parameter of two auxiliary equalization filters (respectively high- and low-pass) which process the noise burst

Finally, SDT puts available an implementation of the continuous crumpling model described in 7.2.3.1.

Practical use of the SDT has been made in the following case studies:

- while simulating impacts between solid surfaces, force data streams provided by the input sensors (see Chapter 2) were pre-conditioned and then analyzed in order to identify foot-floor contact events. Such events have been used to trigger four separate instances of the impact model, corresponding to the heel and toe of each foot: when a contact event was detected, its energy was estimated and the resulting value used to set the initial velocity of the corresponding impact model [225, 226];

- friction has been used while implementing compounds of physically-based building blocks, such as during the synthesis of creaking wood (see Section 7.3);
- concerning soft impacts, the force data stream provided by the FSR sensors (again in Chapter 2) has been treated as in the impact model. In this case however, the energy was used to control the amplitude of the noise burst directly [225, 224, 46, 224];
- as for the continuous crumpling simulations, four pre-conditioned force signals corresponding to the heel and toe of each foot were directly mapped to the respective force parameters of separate instances of the SDT's *crumpling* model. At a lower level, the micro-impacts can be synthesized either by triggering solid or soft impact events. By means of this model, realistic simulations of grounds covered with snow, brushwood or gravel have been obtained [225, 46, 224, 226].

7.2.5 Parametric modeling

In parallel with the synthesis of walking sounds obtained using physically-based models, more simple models can be realized when there is reasonable expectation to come up with realistic, low-latency sonic interactions.

Parametric synthesis has already been proposed for the synthesis of walking sounds [91]. In this section we describe a method that proved effective in the simulation of floors consisting of either homogeneous solid or aggregate surfaces. This method inherits existing analysis-and-synthesis techniques that have been proposed for the generation of walking and, later, hand clapping sounds [65, 234], furthermore it introduces simple novelties in an effort to make the synthesis process as most intuitive for the sonic interaction designer.

The idea is that of starting from the knowledge contained in a pre-recorded set of walking sounds (examples can be downloaded also for free, e.g. from the Freesound online database www.freesound.org). This knowledge is used to i) shape noise by means of linear filters, for instance obtained by LPC processing of the source samples or even by manual tuning of the filter parameters, and then to ii) envelope the amplitude of the filtered noise depending on the instantaneous GRF value. Figure 7.3 shows the method in more detail, as well as the design procedure behind.

Concerning the shaping of the noise source, acceptable results can be obtained by manually tuning a series of second-order bandpass recursive digital filters. Typical parameters are shown in Table 7.1, in which bandwidths are expressed in terms of quality factor (Q) parameters.

In parallel, an amplitude weighing function can be obtained from each material example by aligning and then averaging the envelopes, each obtained through straightforward nonlinear processing [65] of a corresponding source sample. This function ultimately represents a mean envelope signal, whose standard deviation from the average value is known at each temporal step.

At this point one can re-synthesize a sequence of footstep sounds over a different material: when a walking event is detected, the GRF onset (typically the initial 3 or

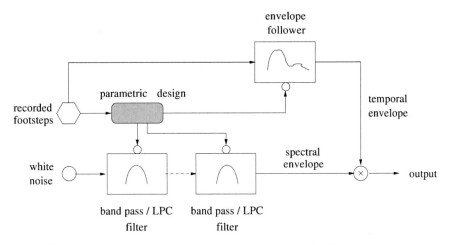

Fig. 7.3: Sketch of the method based on parametric synthesis of footstep sounds.

Material	Filters	Bandwidth (Q)	Center frequency	Gain
Snow	IIR BP	50	400 Hz	1
	IIR BP	700	660 Hz	0.4
Dead leaves	IIR BP	50	100 Hz	1
	IIR BP	500	850 Hz	0.5
	IIR BP	5000	6000 Hz	0.33
Metal (heel)	IIR BP	17	220 Hz	1
	IIR BP	500	220 Hz	1
Metal (toe)	IIR BP	12	250 Hz	1
	IIR BP	20	250 Hz	1
	IIR BP	200	400 Hz	0.24
	IIR BP	500	400 Hz	0.24
Wood (heel)	IIR BP	250	250 Hz	1
	IIR BP	250	250 Hz	1
	IIR BP	180	660 Hz	0.01
	IIR BP	550	660 Hz	0.01
Wood (toe)	IIR BP	55	130 Hz	1
	IIR BP	250	130 Hz	1
	IIR BP	200	610 Hz	0.16
	IIR BP	500	610 Hz	0.16

Table 7.1: Second-order filter parameter values for different ground materials

4 ms) is used to estimate the duration of a footstep. By considering that the energy consumed during every footstep varies to a small extent, the amplitude weighing function is shrunk/stretched along the time dimension (for instance using linear interpolation) and proportionally increased/decreased in amplitude depending on the GRF strength. Furthermore, randomness is added to the mean trajectory at every temporal step proportionally to the standard deviation of the average envelope function.

Parametric synthesis has been successfully adopted to generate accurate stimuli for the test appearing in Section 7.5.2, using linear predictive coding (LPC) to compute the shaping filter coefficients starting from envelope-normalized sound source recordings. Although in that case the procedure was performed offline, the parsimonious use of computational as well as memory resources made by the parametric synthesis procedure poses no problems toward the implementation in real time of this method.

In more detail, the online analysis on the GRF signal is not that simple for parametric synthesis purposes. More elaboration in fact would be needed on top of the onset identification, which keeps into account variations in the foot gesture occurring during the walking act: consider, for instance, a person who stops her foot before completing a step. Until this issues of analysis remain unsolved, physics-based models informed by continuous GRF data deserve more appeal compared to parametric synthesis.

7.3 Composition and parameterization of the models

The sound of a footstep potentially depends on a myriad of contextual conditions: the kind of shoes a subject wears and the type of surface a person is walking on characterize just a subspace of the entire set of possible contexts. One assumption that, for instance, can be made to reduce this space is that the shoe has a rigid sole. This assumption has notable consequences in the simulation of interactions with solid floors, whereas bringing minor effects in the case of aggregate grounds.

The algorithms proposed in the previous sections can be qualitatively informed, starting from a spectral analysis of recordings of real footsteps. From the same recordings, characteristic recurrent events can be identified and then reproduced during the re-synthesis, with an aim toward simulating their evolution across time and subsequently re-combining them into new sounds.

The sound produced while walking on dry leaves can be constructed as a combination of sonic events having relatively long duration and spectral energy at both low and high frequencies, with the addition of noisy glitches that confer "crunchiness" to the final sound. A similar example comes from walking on gravel, resulting from the contribution of physical impacts among stones of different mass, which while colliding give rise to sounds having features that depend on this mass. Such interventions must be correctly parametrized in order to obtain a good match with the components identified in the corresponding real sounds. Finally, an overall loudness must be determined to confer ecological realism to the final sound.

Using different combinations of the models described in the previous section, the following solid and aggregate surfaces have been simulated and tested: wood and creaking wood, metal, deep and low snow, gravel, beach sand, forest underbrush, dry leaves and dirt plus pebbles. [210, 302, 267]. Table 7.2 summarizes how such models have been combined to generate these sounds.

Sound	Impact	Friction	PhISM	Crumpling	GRF control	Stochastic control
Creaking wood	YES	YES	-	-	impacts	friction
Metal	YES	-	-	-	impacts	-
Fresh snow	YES	-	YES	YES	PhISM & crumpling	-
Gravel	-	-	YES	-	PhISM	-
Beach sand	-	-	YES	-	PhISM	-
Forest underbrush	-	-	YES	-	PhISM	-
Dry leaves	-	-	YES	-	PhISM	-
Dirt plus pebbles	-	-	YES	-	PhISM	-

Table 7.2: Combination of sound synthesis models for different walking sounds. Note: for some materials, up to three independent instances of PhISM are computed.

For instance, the sound of footsteps on wood was synthesized by controlling one instance of the SDT impact through GRF data, and superimposing to the impact "creaking" sounds obtained by exciting a friction model through ramp functions driving the external rubbing force. The variation and duration of such ramps was randomly set within certain ranges, before synthesizing every new footstep. The addition of randomness enhances the realism, overall introducing changes in the frequency, amplitude and duration of the sounds.

The examples proposed in this section have been validated through the experiment reported in Section 7.5.1.

7.4 Footstep sounds rendering and displaying

Sounds can be conveyed to the walker through the air and/or bone conduction, by means of speaker, headphone, or contact devices. Concerning display techniques, and depending on the device, approaches can be followed ranging from the traditional mono or stereophonic reproduction up to solutions affording more precise sound source localization, through binaural listening or physical distribution in space of one or more loudspeakers reproducing the virtual source.

Several rendering paradigms have been tested for the synthesis of auditory displays [102]. In recent years, some such paradigms have been challenged in the novel context of interactive display, in which the dynamic evolution of an auditory scene can be obtained by trading off real time constraints and computational burden.

Besides common headphone and room loudspeaker systems, more unusual solutions have been developed for the purpose of realizing ground-level auditory displays: among these, small speakers or insole shakers that are directly mounted into the shoes, and audio-tactile devices placed under the floor. In some cases, walking sounds are delivered by these devices as a by-product of broadband vibrotactile transduction, resulting also in sound as part of the overall reproduced mechanical energy.

7.4.1 Soundscapes rendering and displaying

When exploring a place by walking, at least two categories of sounds can be identified: the person's own footsteps and the surrounding soundscape. In the movie industry, footsteps sounds represent important elements. Chion writes of footstep sounds as being rich in what he refers to as *materializing sound indices*—those features that can lend concreteness and materiality to what is on-screen, or contrarily, make it seem abstracted and unreal [59]. Studies on soundscape originated with the work of R. Murray Schafer [265]. Among other ideas, Schafer proposed sound-walks as empirical methods for identifying a soundscape for a specific location. In a soundwalk people are supposed to move in a specific location, noticing all the environmental sounds heard. Schafer claimed that every place has a sound signature. The idea of experiencing a place by listening has been originally developed by Blesser [41]. By synthesizing technical, aesthetic and humanistic considerations, he describes the field of aural architecture and its importance in everyday life.

For VR purposes, three classes of soundscapes can be identified: static, dynamic, and interactive.

- *Static* soundscapes diffuse an auditory scene regardless of any specific localization effect.
- *Dynamic* soundscapes render the spatial position of one or more sound sources, even dynamically in space, regardless of any user input.
- *Interactive* soundscapes render the auditory scene also as a result of the actions and gestures of the user(s), which for instance can be tracked by a motion capture system. As an example of sound interaction, one can imagine the simulation of a forest, with sounds of fleeing animals following by the movements of a listener furthermore engaged in a walking task.

An engine has been realized in Max/MSP able to provide soundscapes belonging to any of these three classes. To include dynamic and interactive features, the ambisonic tools[4] for Max/MSP were used. Such tools, in fact, allow to move virtual sound sources along trajectories defined on a three-dimensional space [264]. At present, the engine can manage up to sixteen independent virtual sound sources, one to display the user's footsteps and the remaining fifteen handling the external sound sources populating the soundscape.

7.4.2 How to combine footsteps and soundscapes

VR studies made in the field of sound delivery methods and sound quality have recently shown that the addition of environmental cues can lead to measurable enhancement in the sense of presence [280, 61, 263]. Recently, the role of self-produced sound to enhance sense of presence in VE has been investigated. By com-

---

4 Available at http://www.icst.net/research/projects/ambisonics-tools/

bining different kinds of auditory feedback consisting of interactive footstep sounds created by ego-motion with static soundscapes, it was shown how motion in VR is significantly enhanced when moving sound sources and ego-motion are rendered [208].

Specifically in our simulations, a number of soundscapes have been designed according to statistically significant indications given by subjects concerning the sonic ecologies they imagined for a specific environment, e.g. a forest. Such soundscapes were composed mainly by assembling freely available recorded material, like that existing in the Hollywood Edge sound effects library and the Freesound website.

A crucial step for the production of ecologically correct soundscapes consists of balancing the loudness of the background sounds with that of the footsteps, conversely lying in the foreground. This balance was again determined by users during magnitude-adjustment experiments, in which subjects were asked to find out the correct trade-off between the loudness of their own footsteps and the surrounding sounds.

7.5 Evaluating the engines

This section reports on experiments, that were conducted for evaluating the models and techniques described in the previous sections of this chapter.

7.5.1 Auditory recognition of simulated surfaces

The ability of subjects to identify different synthetic ground materials by listening during walking was investigated. In this experiment, subjects were asked to recognize the sounds in an active setting involving microphone acquisition at foot level and subsequent envelope extraction [65] of the subject's walking action. For this reason, the setting was acoustically isolated and subjects were asked to avoid producing sounds other than those generated by their own walking.

7.5.1.1 Methodology and protocol

Sounds were synthesized in real time using the recipes listed in Section 7.3, while subjects were walking across the isolated environment described above.

Participants were exposed to 26 trials, for a total presentation of 13 stimuli each displayed twice in randomized order. The stimuli consisted of footstep sounds on the following surfaces: beach sand, gravel, dirt plus pebbles (like in a country road), snow (in particular deep snow), high grass, forest underbrush (a forest floor composed by dirt, leaves and branches breaking), dry leaves, wood, creaking wood and metal. To increase the ecology of the experiment, footstep sounds on wood, creaking wood and metal were enriched by including some standard room reverberation.

Fifteen participants (six male and nine female), aged between 19 and 29 (mean 22.13, std 2.47), took part in the experiment. All participants reported normal hearing conditions and were naive with respect to the experimental setup and to the purpose of the experiment. They wore sneakers, trainers, boots and other kinds of shoes with rubber sole.

Participants were asked to wear a pair of headphones and to walk in the area delimited by the microphones. They were given a list of different surfaces to be held in one hand, presented as non-forced alternate choice. The list of surfaces presented to the subjects is outlined in the first row of Table 7.3. It represents an extended list of the surfaces the subjects were exposed to.

At the end of the experiment, subjects were asked to answer some questions concerning the naturalness of the interaction with the system. Every participant took on average 24 minutes to complete the experiment.

7.5.1.2 Results

Table 7.3 shows the confusion matrix which displays the results of the experiment. The first row lists the materials that could be chosen, while the first column lists the

	BS	GL	DR	SW	HG	UB	DL	WD	CW	MT	WR	CR	MR	FS	CC	PD	WT	CP	—
BS	15	2		5			1							2				1	4
GL		21	2			1	1							4					1
DR		1	3	2		6	6	1						10					1
SW				24	1									4					1
HG	2	7	3	1	0	3	7			2									5
UB		1	3	1		19	1							3			1		1
DL	1	3	5			5	12							4					
WD			1	2				14		1						1	1		10
CW								1	28			1							
MT								1		24		1			2				2
WR								3		11	6			7					3
CR												28		1					1
MR								1					25	1	1				2

Abbreviations:

WD wood	CW creaking wood	SW snow	UB underbrush
— don't know	FS Frozen snow	BS beach sand	GL Gravel
MT metal	HG High grass	DL dry leaves	CC concrete
DR dirt	PD puddles	WT Water	CP carpet
WR wood reverb	MR metal reverb	CR creaking+ reverb	

Table 7.3: Confusion matrix: recognition of synthesized footstep sounds.

materials simulated in the stimuli subjects were exposed to. The decision of providing a wider choice of materials was taken to minimize the statistical significance of subjects guessing at random.

From this table, it is possible to notice how surfaces such as snow, creaking wood with and without reverberation, gravel and metal with reverberation were correctly recognized in a high number of trials. Recognition of surfaces such as dirt plus pebbles, high grass and wood appeared to be low. An analysis performed on the wrong answers reveals that on average subjects tended to mistakenly spread judgments over surfaces belonging to the same category (e.g., wood versus concrete, snow versus frozen snow, dry leaves versus forest underbrush) while keeping different categories distinct in their judgments (e.g., wood versus water, wood versus gravel, metal versus dry leaves).

Moreover, results show that the addition of reverberation to the sounds resulted in better recognitions for metal, and worse for wood, which was perceived most of the times as concrete. Overall, recognition rates are similar to those measured on recorded footstep sounds [210].

7.5.2 Salience of temporal and spectral cues of walking

Based on the parametric synthesis model described in Section 7.2.5, an experiment has been performed aiming to understand the salience of auditory cues of walking.

The experimental hypothesis was kept simple, by relying on a classification of such cues in spectral and temporal. Furthermore the experiment itself was made offline, in this way focusing on the auditory feedback alone while excluding vision and touch. The complete report on this activity can be found in [94].

We hypothesized that the perception of solid materials is mainly determined by *spectral* cues, conversely the perception of aggregate materials is mainly determined by *temporal* cues. In particular, we experimented using concrete (C) and wooden (W) floors, representative of solid materials, as well as with gravel (G) and dried twigs (T), representative of aggregate materials. The former, such as concrete, marble, wood, are stiff. The latter, such as gravel, dry leaves, sand, allow relative motion of their constituent units and progressively adapt to the sole profile during the interaction. Now,

- solid materials give rise to short, repeatable impacts having a definite spectral color;
- aggregate materials elicit sequences of tiny impacts having distinctive temporal density, that create a sort of "crumpling", less resonant sound.

Figure 7.4 illustrates the hypothesis.

MATERIAL	PHYSICAL PROPERTIES	ACOUSTIC PROPERTIES
C , W	Solid	Spectral Cues
G , T	Aggregate	Temporal Cues

Fig. 7.4: Experimental hypothesis. (C: concrete, W: wood, G: gravel, T: twigs.)

7.5.2.1 Methodology and Protocol

Subjects were sitting in front of a Mac Pro PC running a Java application communicating (via the *pdj* library) with Puredata, a free software environment for real time audio synthesis also enabling simple visualizations (through the *GEM* library). They listened to the auditory stimuli through a pair of AKG K240 headphones.

Thirteen male and three female undergraduate computer science students aged aged 22 to 31 (mean = 24.62, std = 2.55) participated in the experiment. Few of them had some experience in sound processing. All of them reported to usually wear snickers.

At the end of the experiment, every subject completed a subjective questionnaire about the realism and ease of identification of the audio stimuli.

One footstep by a normally walking male wearing leather shoes was repeatedly recorded while he stepped over a tray filled with gravel and, then, dried twigs. Recordings were made inside a silent, normally reverberant room using a Zoom H2 digital hand recorder standing 0.5 m far from the tray. For either material, seven recordings were selected and randomly enqueued to create walking sequences lasting 12 s and containing 13 footsteps. In addition to the in-house recordings, high quality samples of a male walking on concrete and on wooden parquet were downloaded from the commercial database `sounddogs.com`. Using these samples, two further walking sequences were created having the same beat and average Sound Pressure Level as of those based on in-house recordings.

Temporal envelopes were extracted from every sequence, by computing the signal

$$e_M[n] = (1 - b[n])|s_M[n]| + b[n]e_M[n-1] \tag{7.10}$$

out of the corresponding sequence s_M, $M \in \mathcal{M} = \{C, W, G, T\}$. (Refer to Figure 7.4 for the meaning of the C, W, G, and T.) As in previous research on synthetic footsteps, the envelope following parameter $b[n]$ was set to 0.8 when $|s_M[n]| > e_M[n-1]$, and to 0.998 otherwise [65]. By following the input when its magnitude is greater than the envelope, and by in parallel allowing a comparably slow decay of the envelope itself when the same magnitude is smaller, this setting ensures that amplitude peaks are tracked accurately, while leaving spurious peaking components off the envelope signal e_M.

By dividing every sequence s_M by its envelope e_M, we computed signals $u_M = s_M/e_M$ in which the temporal dynamics was removed. In other words, we manipulated the footstep sequences so to have stationary amplitude along time.

What remained in u_M was a spectral color, that we extracted with a 48th-order inverse LPC filter h_M^{-1} estimated in correspondence of those parts of the signals containing footstep sounds. Using this filter order, if training the model using *one* footstep then we could not detect differences between the original sound and the correspondingly re-synthesized footstep. We emphasize that the resulting LPC filter in any case estimated one single transfer function, independently of the number of footsteps taken from the original sequence which informed the model. Since we trained the estimator with the entire sequence, the re-synthesized sound had a slightly different color compared to any other footstep belonging to the original sequence.

In the end, for every material M a highly realistic version $\tilde{s}_M$ of the original sequence s_M could be re-synthesized by convolving digital white noise w by the "coloring" filter h_M, and then multiplying its output, i.e. the synthetic version $\tilde{u}_M$ of u_M, by the envelope signal e_M:

$$\tilde{s}_M[n] = (w * h_M)[n] \cdot e_M[n] = \tilde{u}_M[n] \cdot e_M[n]. \tag{7.11}$$

This technique draws ideas from a family of physically-informed models of walking sounds [65, 303]. In the meantime it provides a simpler, more controlled re-synthesis process avoiding stochastic generation of patterns as in such models. In our case, the silent parts of the four envelopes were tailored to generate synthetic sequences having identical walking tempos. This simple manipulation ensured seamless mutual exchange of the envelopes among sequences, as explained in the following.

Sixteen stimuli were created by adding twelve *hybrid* re-syntheses to the *native* stimuli $\tilde{s}_C$, $\tilde{s}_W$, $\tilde{s}_G$, and $\tilde{s}_T$. Every hybrid stimulus $\tilde{s}_{M_t,M_f}$, $M_t, M_f \in \mathcal{M}$ was defined as to account for the spectral color of material M_f *and* the temporal envelope of material $M_t \neq M_f$:

$$\tilde{s}_{M_t,M_f}[n] = (w * h_{M_f})[n] \cdot e_{M_t}[n] = \tilde{u}_{M_f}[n] \cdot e_{M_t}[n]. \tag{7.12}$$

For each material M_f, we checked that all hybrid temporal manipulations using $M_t \neq M_f$ did not notably alter the spectral information of $\tilde{s}_{M_f}$, and thus its original color. In fact, an inspection of the spectra $E_M(\omega)$ of the various envelopes made by Fourier-transforming e_M, i.e., $E_M(\omega) = \mathscr{F}\{e_M\}(\omega)$, shows that they all have a comparable spectrum. More precisely, all spectra E_C, E_W, E_G, E_T exhibit similar magnitudes, that are shown in Figure 7.5 after removing the respective dc component for ease of inspection. This means that the spectral differences in $\tilde{S}_{M_t,M_f}(\omega)$ caused by multiplying $\tilde{u}_{M_f}$ by e_{M_t}, that is,

$$\tilde{S}_{M_t,M_f}(\omega) = \mathscr{F}\{\tilde{u}_{M_f} \cdot e_{M_t}\}(\omega) = (\tilde{U}_{M_f} * E_{M_t})(\omega), \tag{7.13}$$

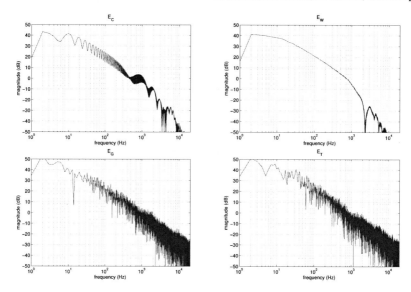

Fig. 7.5: Magnitude spectra of envelopes E_C, E_W, E_G, and E_T. Respective dc components removed for ease of inspection.

are substantially independent of the material, hence almost identical to those introduced in $\tilde{s}_{M_f}$ by its own envelope e_{M_f}.

Symmetrically, the temporal artifacts which are caused by hybridization between two different materials can be considered minor. In fact, because of the LPC design methodology, all filters h_C, h_W, h_G, h_T do transform white noise into a stationary signal independently of the material.

To become confident with the auditory stimuli, subjects trained for some minutes before starting an individual session by selecting and playing each one of the original sequences s_C, s_W, s_G, s_T for several times. Each sequence could be selected by clicking on the corresponding software button in a graphic interface.

Each individual session consisted of 192 trials, obtained by randomly playing each one of the sixteen synthetic stimuli for twelve times. At each trial the subject listened to a stimulus, and selected one material by clicking the corresponding button in the interface. When the button was released, the screen froze for two seconds and changed color to inform subjects of the conclusion of the trial. After this short pause, a new trial was performed.

The four buttons randomly switched position at each trial. Subjects could temporarily stop the experiment by clicking the pause icon '||' located in the middle of the screen, whenever they wanted to take a short break among trials. It took approximately 45 minutes for each participant to complete the session.

7.5.2.2 Results

For each participant, percentages of selection for the four materials C, W, G, T were analyzed. We considered the fraction of participants who showed significantly correct (random is 25%) material recognitions from the four synthetic stimuli $\tilde{s}_C$, $\tilde{s}_W$, $\tilde{s}_G$, $\tilde{s}_T$, across the twelve repetitions. The critical value (with $\alpha = 0.05$) of the one-tailed binomial test Bin(12,0.25) is 7 trials (i.e., 58.33%): only the participants with an auditory recognition of the original materials higher than 58.33% were included in the analysis. After this check, 16 participants were considered for the recognition of dried twigs, 15 for gravel, 16 for wood, and 10 for concrete.

The results of the analysis are presented in Figure 7.6. In these plots, a bar exhibiting a low percentage means that the correspondingly substituted information (either temporal or spatial) is important for the recognition of the original material, represented by the leftmost bar in the same plot. The difference from random percentage (25%) was tested using one-proportion (two-tailed) z tests.

By aggregating the data, we also evaluated the auditory recognition of material categories. Again, this analysis was conducted using data from participants exhibiting an auditory recognition significantly higher than random concerning the two sets of stimuli accounting for the respective categories (24 trials for each category). In this case, the critical value (this time computed by a one-proportion/one-tailed z test to account for the larger number of trials, with $\alpha = 0.05$) is 10 trials, corresponding to 41.67%. All the participants passed the check.

The results of the new analysis are presented in Figure 7.7. For the different percentages of selection, the difference relative to random (25%) was tested using one-proportion (two-tailed) z tests. Thus, for the aggregate category, the percentages of selection in native (82.29%) and frequency manipulated (52.78%) conditions were significantly different from random ($z = 25.98$, $p < 0.001$ and $z = 21.77$, $p < 0.001$, respectively). By contrast, the percentage of selection in the time manipulated condition (23.44%) was not significantly different from random ($z = -1.22$, $p = 0.22$). On the other hand, for the solid category, the percentages of selection in native (78.39%) and frequency manipulated (35.59%) conditions were significantly different from random ($z = 24.16$, $p < 0.001$ and $z = 8.30$, $p < 0.001$, respectively). By contrast, the percentage of selection in the time manipulated condition (26.65%) was not significantly different from random ($z = 1.29$, $p = 0.20$). The differences between the three audio conditions were tested with two-proportion (two-tailed) z tests.

A correction for experiment-wise error was realized by using Bonferroni-adjusted alpha level (p divided by the number of tests). Thus, in order to compare the three audio conditions (native, frequency manipulated, and time manipulated), the alpha level was adjusted to $p = 0.05/3 = 0.0167$. For the aggregate category, the analysis showed that the native condition was significantly different from the frequency manipulated ($z = 10.23$, $p < 0.05$) and time manipulated ($z = 20.56$, $p < 0.05$) conditions. The difference between frequency manipulated and time manipulated conditions was significantly different ($z = 14.50$, $p < 0.05$) as well. For the solid category, the analysis indicated that the native condition was significantly different from

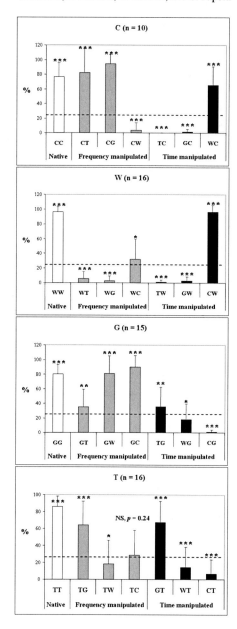

Fig. 7.6: Mean percentages of selection (lines represent std) for C,W,G,T as a function of the auditory stimulus $\bar{s}_{M_t,M_f}$. The difference from random selection (line at 25%) was tested using one-proportion (two-tailed) z tests. Note: $*: p < 0.05$, $**: p < 0.01$, $***: p < 0.001$, NS: not significant, n: number of subjects.

the frequency manipulated ($z = 14.57$, $p < 0.05$) and time manipulated ($z = 17.96$, $p < 0.05$) conditions. The difference between frequency manipulated and time manipulated conditions was also significantly different ($z = 4.63$, $p < 0.05$).

After the experiment, a questionnaire was proposed in which each participant had to grade from 1 to 7 the four native stimuli according to two subjective crite-

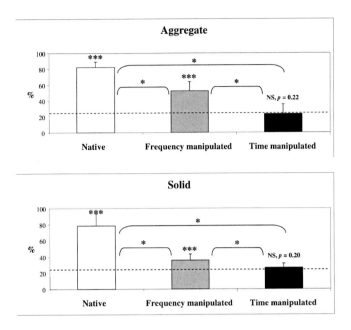

Fig. 7.7: Mean percentages of selection (lines represent std) for material categories (Aggregate and Solid) as a function of the auditory stimulus $\tilde{s}_{M_t,M_f}$. The difference from random selection (line at 25%) was tested using one-proportion (two-tailed) z tests. The differences between the three audio conditions were tested with two-proportion z tests (two-tailed and Bonferroni-adjusted alpha level with $p = 0.05/3 = 0.0167$). Note: $*: p < 0.05$, $**: p < 0.01$, $***: p < 0.001$, NS: not significant.

ria: realism, and ease of identification. Wilcoxon signed rank (two-tailed) tests with Bonferroni correction showed significant differences only for the realism of sounds: between concrete and dried twigs ($z = -3.28$, $p = 0.001$), and between concrete and gravel ($z = -3.16$, $p = 0.0016$).

7.5.2.3 Discussion and Conclusions

The histograms for concrete and wood in Figure 7.6 show that subjects tolerate swapping between the temporal features of C and W, both belonging to the solid category, conversely the substitution in the same signals with temporal features extracted from aggregate materials (i.e. G and T) harms the recognition. This result is in favor of the initial hypothesis. The effect of spectral manipulations of C and W is more articulate. In this case the hypothesis is essentially confirmed with wood, whose distinctive color cannot be changed using any other spectrum. In parallel, subjects are tolerant to substitutions in C with spectra from aggregate materials. This greater tolerance may be due to the basic lack of distinct color of concrete floors, especially for listeners who usually wear rubber sole shoes such as snickers (indeed the majority of our sample). The same conclusion finds partial confirmation by the greater confidence shown by subjects in recognizing aggregate materials, in the limits of the significance of these data.

The histograms in Figure 7.6 regarding gravel and twigs partially support the initial hypothesis. Time swaps between G and T are tolerated to a lesser extent compared to solid floors. Like before, substituting the temporal features of solid materials in an aggregate sound is not tolerated. Spectral substitutions are not as destructive as they were for solid materials, especially in the case of gravel. The worst situation is when the spectrum of W is substituted in T, again probably due to the distinct color that wood resonances bring into the sound.

Figure 7.7 would further support this discussion. In fact, in spite of the low significance of the data from time manipulations (i.e. black bars), it shows that subjects are primarily sensitive to temporal substitutions between solid and aggregate materials. In parallel, spectral changes are more tolerated during the recognition of aggregate material compared to solid floors.

The proposed experiment has confirmed that solid and aggregate floor materials exhibit precise temporal features, that cannot be interchanged while designing accurate walking sounds. Within such respective categories, spectral color represents an important cue for the recognition of solid materials, conversely sounds of aggregate materials seem to tolerate larger artefacts in their spectra.

7.5.3 Evaluation of soundscapes: interactivity

In two preliminary experiments [300] it was investigated how subjects react to different soundscape dynamics while walking in a virtual auditory scene.

The task of the first experiment was to walk across a circular perimeter inside the walking area in a room. Eight loudspeakers were placed at the angles and middle points of each side of a rectangular floor. The user's position was tracked by a Mo-Cap system, and then used to localize synthetic footsteps along the task through the speakers. Localization was performed using the ambisonics tools for Max/MSP, allowing to move virtual sound sources on a three-dimensional space (refer to Section 7.4.1). During the experiment the loudspeakers were hidden by opaque, acoustically transparent curtains.

During the walk, subjects were exposed to six conditions—again, refer to Section 7.4.1: i) static soundscape; ii) coherent interactive soundscape; iii) incoherent interactive soundscape; iv) static soundscape with static distractors; v) coherent interactive soundscape with dynamic distractors; vi) incoherent interactive soundscape with dynamic distractors. Incoherent means that the footstep sounds were localized opposite to the actual position of the moving subject in the walking area.

Distractors consisted of footstep sounds of a phantom subject walking in the same area. Specifically, static distractors were rendered by displaying their sound with equal intensity from all room loudspeakers, whereas dynamic distractors covered a triangular trajectory inside the circular perimeter.

Participants were exposed to twelve trials, where the six conditions were presented twice in randomized order. Performing each trial took about one minute. Each condition was presented using virtual floors made of wood and forest underbrush. The reason for choosing one solid and one aggregate material was to discern whether the surface type affected the results.

The distractors displayed the same virtual surfaces, but at lower intensity, slightly different timbre, and moderately faster gait cycle.

After the presentation of each stimulus, participants were required to answer the following questions on a seven-point Likert scale:

- How well could you localize under your feet the footstep sounds you produced?
- How well did the sounds of your footsteps follow your position in the room?
- How much did your walk in the VE seem consistent with your walk in the real world?
- How natural did your interaction with the environment seem?
- To what degree did you feel confused or disoriented while walking?

Our hypotheses were that a coherent interactive soundscape would result in improved subjective appreciation and consequent rating; that the incoherent dynamic condition would have been judged as the worst; and that the use of distractors would have decreased the subjective appreciation of the perceived auditory scene, hence its rating.

First, a significant difference was found between the surface materials for what concerns the coherent interactive and static condition, in the case of absence of distractors: the difference between such conditions is negligible in the case of wood, whereas this difference becomes significant in the case of forest underbrush ($p < 0.0001$). Conversely, the same difference was not significant in presence of the distractors.

Secondly, for both materials the incoherent interactive condition gave rise to poorer evaluations in terms of localization, following, consistency and naturalness as well as to less confidence on orientation, both in presence and in absence of distractors. In detail, for both materials significance was found concerning the difference between the coherent and incoherent interactive conditions (for wood: $p < 0.001$ and $p < 0.01$; for forest: $p < 0.000001$ and $p < 0.05$, both respectively with and without distractors), as well as between the static and incoherent interactive conditions (for wood: $p < 0.01$ and $p < 0.01$; for forest: $p < 0.000001$, both respectively with and without distractors).

Thirdly, for both materials the evaluations in absence of distractors were almost always better than in presence of them concerning localization, following, consistency, naturalness and orientation. This difference was significant for the forest underbrush case ($p < 0.05$), whereas for the wood it was not.

For both materials, the disorientation was higher in presence rather than in absence of distractors, but significant differences between these two conditions were found only for the forest underbrush case ($p < 0.05$). The incoherent interactive with distractors condition was rated as the most disorienting for both materials.

Conversely, for the forest underbrush case the coherent interactive condition was rated as the least disorienting.

Overall, the coherent interactive condition gave rise to significantly better results than the static one concerning the forest underbrush case in absence of distractors. A subsequent analysis for each of the investigated parameters revealed significant differences between the two conditions only for the "naturalness" parameter ($p < 0.05$).

It is therefore possible to conclude that users can perceive that their interaction with the VE is neither realistic nor natural when the source is not moving coherently with their position. The hypothesis concerning the distractors was confirmed: for both materials, almost always the evaluations in absence of distractors were better than when the distractors were present, although significant differences were found only for the forest case. In addition, the disorientation was higher in presence of distractors (but significant only for the forest case). This evidence suggests that the use of distractors, i.e., walking sounds evoking the presence of another person walking in the same room as the subject, is likely to influence the perception of self-produced footstep sounds.

Starting from the results of the first experiment we designed a second experiment, investigating in more detail the subjective perception of the static and coherent interactive soundscapes. The task consisted of walking freely inside the walking area. Participants were exposed to fourteen trials, where seven virtual surface materials were randomly presented in presence of both types of soundscapes. Such materials, five aggregate and two solid, consisted of gravel, sand, snow, dry leaves, forest underbrush, wood and metal.

Each trial lasted about one minute. After the presentation of each stimulus participants were required to evaluate, on a seven-point Likert scale, the same questions presented in the first experiment.

The goal of this experiment was to assess whether participants showed a preference for either display method, while exploring the VE during a free walk (i.e., without any predefined trajectory like in the first experiment). Furthermore we were interested in assessing whether the surface property affected the results.

Results show that participants did not show any preference for either method. The answers to the questionnaire were very similar for all the surfaces, with no significant differences. This result suggests that both methods could be used in a VE, to deliver interactively generated footstep sounds. However, other tests should be conducted to add quantitative elements to this conclusion.

7.5.4 Evaluation of soundscapes: ecology

An experiment was conducted aiming at understanding the role of soundscapes in creating a sense of place and context when designing a virtual walking experience [300]. More in detail, the goal of the experiment was to investigate the ability of subjects to recognize the different walking sounds they were exposed to in three conditions: without soundscape, with ecologically coherent soundscape (e.g. footsteps sounds on a soundscape reporting of a beach) and with ecologically incoherent soundscape (e.g. footsteps sounds on a soundscape reporting of a ski slope).

The interactive footsteps were synthesized in real time while subjects were walking using the system described in Section 7.4.1. Offline, the following soundscapes were built: a crowded beach, the courtyard of a farm, a ski slope, a forest, and a garden with trees during fall. All soundscapes were diffused as static.

The task was to recognize the surface material, as well as to evaluate the realism and quality of the footstep sounds. In the conditions with soundscape, participants were also asked to recognize the surrounding environment in which they were walking.

Results showed that the addition of a coherent soundscape resulted in a better recognition of the surfaces, along with a higher realism and quality of the proposed sound compared to the conditions without and with incoherent soundscape.

For some surface materials, adding a coherent soundscape significantly improved the surface recognition compared to the case in which the soundscape was not provided, and this happened especially with materials whose recognition was difficult without soundscape. Similarly, the percentages of correct answers were higher in the condition with coherent soundscape compared to the condition with incoherent soundscape, significantly for some materials. Furthermore, the same percentages were higher in the condition without soundscape compared to the condition with incoherent soundscape. As expected, adding an incoherent soundscape created an ecological mismatch which often confused the subjects.

The analysis of the wrong answers reveals that in all the experiments none of the proposed aggregate surfaces was confused with a solid one. This means that subjects were able to robustly identify the type of surface.

Regarding the evaluations on the realism and quality of the footsteps sounds in the three conditions, higher evaluations were found in the condition with coherent soundscape compared to the condition without and with incoherent soundscape, as well as for the condition without soundscape compared to that with incoherent soundscape. For some materials these evaluations were statistically significant.

Additionally, the percentages of correct guess of the soundscape were higher with coherent rather than incoherent soundscape.

Overall, subjects observed that soundscapes play an important role in ground surface recognition, precisely for their ability to create a context. Especially in presence of conflicting information, as it was the case with incoherent soundscapes, subjects tried to identify the strongest ecological cues in the auditory scene while performing their recognitions.

This experiment gives strong indication of the importance of context in the recognition of a virtual auditory scene, where walking sounds generated by subjects and soundscapes are combined. Though, it is only a preliminary investigation: further experiments are needed to gain a better understanding of the cognitive factors involved when subjects are exposed to different sound events, especially when a situation of semantic incongruence is present.

7.6 Conclusions

This chapter provided a description of how to synthesize walking sounds using physics-based and physically inspired models, including recipes for constructing and parameterizing model compositions dictated by the designer's experience and taste. Several surfaces have been simulated, both solid and aggregate. The simulations work in real time and are controlled by kinds of input devices such as those described in Chapter 2.

After reporting on current auditory display possibilities, we also described experiments whose aim was to assess the ability of subjects to recognize the simulated surfaces, the saliency of temporal and frequency cues in footstep sounds, the role of soundscapes in enhancing the interactivity and ecological realism of an auditory scene. These experiments validate the quality of the proposed synthesis engines, and testify their possibilities and limits to faithfully recreate virtual walking experiences.

Chapter 8
Multisensory and haptic rendering of complex virtual grounds

G. Cirio, Y. Visell, and M. Marchal

Abstract The addition of vibrotactile and, more generally, multimodal feedback when interacting with a virtual environment is fundamental when aiming at fully immersive and realistic simulations. This is particularly true when using natural navigation paradigms such as walking for the exploration of virtual environments. Tactile perceptual cues generated by a ground surface can provide crucial information regarding the ground material itself, the surrounding environment and specificities of the foot-floor interaction, such as gait phase or forces, and can even reflect user emotions. This chapter addresses the multimodal rendering of walking interactions with virtual ground surfaces, incorporating vibrotactile, acoustic and graphic rendering to enable truly multimodal experiences. Taking advantage of the availability of novel multimodal floor surfaces (see Chapters 2 and 3), we propose different models for the rendering of vibrotactile and multimodal cues from the foot-based interaction with two categories of complex ground materials that exhibit strong high-frequency components: granular materials and fluids.

8.1 Context

VR applications aim at simulating digital environments with which users can interact and, as a result, perceive through different modalities the effects of their actions in real time. In real life, we interact with our surrounding environment with our five senses. Ideally, this should also be the case in a VR simulation. However, most VR simulators built to date contain visual displays, vision-based or mechanical tracking devices to monitor the position of users and props, and spatialized sound displays, but neglect the haptic sense, or sense of touch, not to mention those of olfaction (smell) and gustation (taste). Consequently, current typical VR applications rely primarily on vision and hearing.

Because touching is an integral part of our experience of the world, in order to improve the immersion of users in virtual environments, haptic (force and/or tactile) feedback is essential, to enable users to touch, feel, and manipulate objects. Virtually all tasks we accomplish in real life involve bodily interaction with the environment. It also appears to be true that a higher sense of presence can be achieved in a VR simulation through the addition of even low-fidelity tactile feedback to an existing visual and auditory display than can be accomplished by improving one particular modality such as the visual display alone [275].

Just as the synthesizing and rendering of visual images defines the area of computer graphics, the art and science of developing devices and algorithms that synthesize computer generated force-feedback and vibrotactile cues is the concern of computer haptics [29]. Haptics broadly refers to touch interactions (physical contact) that occur for the purpose of perception or manipulation of objects [262].

During walking interactions, low-frequency forces due to movements of a walker's lower body generate low-frequency ground reaction forces as well as higher frequency acoustic and vibrotactile signals due to foot interaction with the ground material. In this chapter, we focus on the high frequency components of mechanical signals generated during walking interactions, which are readily reproduced via relatively low cost vibrotactile display devices or by auditory displays. Specifically, we study the the modeling and simulation of different ground materials allowing the multimodal rendering of foot-ground interaction, with special emphasis on the vibrotactile modality, but also incorporating the acoustic and visual sensory channels.

8.1.1 Vibrotactile rendering of virtual materials

Tactile rendering refers to the process by which sensory stimuli are computed using a software algorithm in order to convey information about a virtual object through the tactile modality. Vibrotactile displays primarily address the human haptic ability to perceive high-frequency mechanical stimuli, with frequency content distributed primarily between 30 Hz and 800 Hz. Vibrotactile rendering algorithms gather data from the environment, such as the feet position and the physical attributes of the virtual objects (shape, elasticity, texture, mass, etc), and compute the vibrotactile signals that would result from these interactions. The design of the rendering algorithm is crucial for an accurate stimuli restitution: a simple pre-computed signal (e.g., the playback of a pre-recorded impact force transient signal) is likely to feel different than a more complex signal generated using a physically-based model, since the latter can vary depending on the physical parameters governing the interaction.

Transient signals have been used at the moment of impact to improve the perception of contact with rigid bodies of different material. Okamura et al. [215] recorded real, high resolution acceleration information by tapping on different materials with a measuring instrument. This data was fit to decaying sinusoidal signals of the form $S(t) = A(v) \exp{-Bt} \sin(\omega t)$, where $A(v)$ is the amplitude depending on the tool velocity, B is a decay constant, and ω the oscillation frequency. Therefore, impacts with each captured material was represented by a different vibration signature. This vibrotactile signal was then rendered by an appropriate vibrotactile transducer at the moment of impact, successfully conveying perceptual information about the material hardness.

However, transducers do not exactly reproduce what is recorded through other devices. Hence, Okamura's technique was later improved [214] to compensate for device dynamics and rendering bandwidth by applying scaling factors found through human perceptual experiments for a given transducer. This lead to the rendering of realistic signals corresponding to materials such as rubber, wood and aluminum, but had the main drawback of requiring a perceptual tuning step for each transducer. Kuchenbecker et al. [153] addressed this issue by adapting the signal to the dynamic response of the device using an inverted system model of the display. This enabled the rendering of realistic impacts from recorded force patterns. Further studies by Fiene et al. improved this technique by considering the grip force applied to the device [92].

8.1.2 Contributions

In the context of natural walking, prior work has not addressed the vibrotactile rendering of walking interactions with virtual grounds. Yet, including tactile cues when exploring virtual environments would bring major benefits in the fields of medical rehabilitation for gait and postural exercises, training simulations for the recreation of compelling and realistic grounds, and entertainment for improved immersion within rich virtual environments.

Taking advantage of the availability of novel multimodal floor surfaces (see Chapters 2 and 3), we propose different models for the rendering of vibrotactile and multimodal cues from the foot-based interaction with two categories of complex ground materials that exhibit strong high-frequency components: granular materials and fluids.

Footsteps onto granular (aggregate) ground materials, such as sand, snow, or ice fragments belie a common temporal process originating with the transition toward a minimum-energy configuration of an ensemble of microscopic systems, via a sequence of transient events. The latter are characterized by energies and transition times that depend on the characteristics of the system and the amount of power it absorbs while changing configuration. They dynamically capture macroscopic information about the resulting composite system through time. On the other hand, liquid-covered ground surfaces, such as water puddles and shallow pools, have an important kinesthetic component due to pressure and viscosity forces within the fluid, and may, at first, seem to lack high frequency mechanical responses. However, important high frequency components exist, as generated by bubble and air cavity resonances, which are responsible for the characteristic sound of moving fluids. We utilize the fact that vibrotactile and acoustic phenomena share a common physical source by designing our vibrotactile models based on existing knowledge of fluid sound rendering. Both types of ground materials exhibit very interesting high frequency features adequate for their restitution through an actuated vibrotactile floor: as opposed to rigid surfaces, the overall signal is not reduced to transients at the moment of impact, but can produce a signal during the entire foot-floor contact duration.

Although mainly focused on the vibrotactile modality, our approaches are multimodal. The same models can be used to synthesize acoustic feedback, due to vibrotactile and acoustic phenomena common generation mechanisms and physical source. The visual modality is an absolute requirement on its own, since interacting with virtual environments without visual feedback is of little interest, except in very specific cases.

8.2 Walking on disordered natural materials

In this section, we present techniques to enable users to interact on foot with simulated natural ground surfaces, such as soil or ice, in immersive virtual environments, using the interface and interaction techniques described in Chapter 3. Position and force estimates from in-floor force sensors are used to synthesize plausible auditory and vibrotactile feedback in response. Relevant rendering techniques are discussed in the context of specific interactive scenarios, involving walking on a virtual frozen pond or bed of sand.

Sensations accompanying walking on natural ground surfaces in real world environments (sand in the desert, or snow in winter) are multimodal and highly evocative of the settings in which they occur [309]. Limited prior research has addressed foot-based interaction with virtual and augmented reality environments [159, 309], perhaps due to a lack of efficient techniques for capturing foot-floor contact interactions and rendering them over distributed floor areas, and the emerging nature of the applications involved.

Here, we present a novel solution using a network of instrumented floor tiles, and methods for capturing foot-floor contact interactions so as to render multimodal responses of virtual ground materials.

8.2.1 Vibrotactile rendering of stepping on disordered heterogeneous materials

Due to the highly interactive nature of the generation of haptic stimuli in response to foot-applied pressure, the display of haptic textures, in the form of high frequency vibrations simulating the feel of stepping onto heterogeneous solid ground materials [308], is a significant challenge to be overcome in the multimodal rendering of walking on virtual ground surfaces. During a step onto quasi-brittle porous natural materials (e.g., sand or gravel), one evokes physical interaction forces that include viscoelastic components, describing the recoverable deformation of the volume of the ground surrounding the contact interface; transient shocks from the impact of

foot against the ground; and plastic components from the collapse of brittle structures or granular force chains, resulting in unrecoverable deformations [78, 279].
Combinations of such effects give rise to the high frequency, texture-like vibrations
characteristic of the feel of walking on different surfaces [86]. Figure 8.1 presents
an example of force and vibration data acquired by the authors from one footstep
on a gravel surface. Because the vibration signature is continuously coupled to the

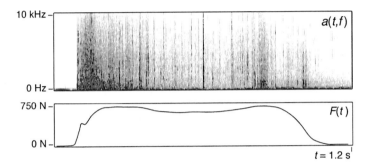

Fig. 8.1: Vibration spectrogram $a(t, f)$ and normal force $F(t)$ measured from one footstep onto
rock gravel (Authors' recording). Note the discrete (impulsive) broadband impact events evidenced
by vertical lines in the spectrogram.

force input over time in such examples, there is no straightforward way to con-
vincingly use recorded footstep vibrations for vibrotactile rendering, although more
flexible granular sound-synthesis methods could be used [28, 69]. For the modeling
of simpler interactions, involving impulsive contact with solid materials, recorded
transient playback techniques could be used [153].

A simple yet physically-motivated approach we have taken to the haptic synthesis
of interaction with such surfaces is based on a minimal fracture mechanics model,
drawing on an approach that has also proved useful for modeling other types of
haptic interaction involving damage [112, 179]. Figure 8.2 illustrates the continuum
model and a simple mechanical analog used for synthesis.

In the stuck state, the surface has stiffness $K = k_1 + k_2$, effective mass m and
damping constant b. It undergoes a displacement x in response to a force F, as
governed by:

$$F(t) = m\ddot{x} + b\dot{x} + K(x - x_0), \quad x_0 = k_2 \xi(t)/K \tag{8.1}$$

In the stuck state, virtual surface admittance $Y(s) = \dot{x}(s)/F(s)$ is given, in the
Laplace-transformed (s) domain, by:

$$Y(s) = s(ms^2 + bs + K)^{-1}, \quad K = k_1 k_2 \xi/(k_1 + k_2) \tag{8.2}$$

where $\xi(t)$ represents the net plastic displacement up to time t. A Mohr-Coulomb
yield criterion is applied to determine slip onset: When the force on the plastic unit
exceeds a threshold value (which may be constant or noise-dependent), a slip event

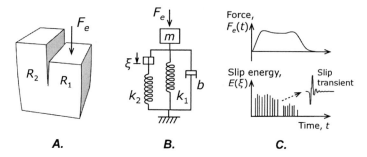

Fig. 8.2: Normal force texture synthesis. *A.* A fracture mechanics approach is adopted. A visco-elasto-plastic body undergoes shear sliding fracture due to applied force F_e. *B.* A simple mechanical analog for the generation of slip events $\xi(t)$ in response to F_e. *C.* For vibrotactile display, each slip event is rendered as an impulsive transient using an event-based approach.

generates an incremental displacement $\Delta\xi(t)$, along with an energy loss of ΔW representing the inelastic work of fracture growth.

Slip displacements are rendered as discrete transient signals, using an event-based approach [153]. High frequency components of such transient mechanical events are known to depend in detail on the materials and forces of interaction, and we model some of these dependencies when synthesizing the transients [312]. An example normal force texture resulting from a footstep load during walking is shown in Figure 8.3).

Fig. 8.3: Example footstep normal force and synthe-sized waveform using the simple normal force texture algorithm described in the text. The respective signals were captured through force and acceleration sensors in-tegrated in the vibrotactile display device described in Chapters 2 and 3.

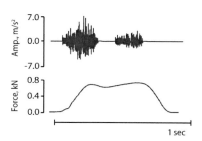

8.2.2 Multimodal rendering of material interactions with a virtual frozen pond

Based on the interface and interaction techniques presented in Chapter 3, we de-signed a virtual frozen pond demonstration that users may walk on, producing pat-terns of surface cracks that are rendered and displayed via audio, visual, and vibrotactile channels (Figure 8.4). The advantage of this scenario is that plausibly realistic

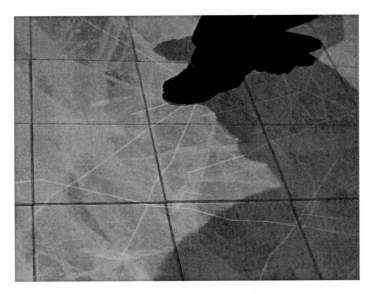

Fig. 8.4: Still images of users interacting with the simulated frozen pond.

visual feedback could be rendered without detailed knowledge of foot-floor contact conditions, which would require a more complex sensing configuration.

8.2.2.1 Non-visual rendering

In the demonstration, audio and vibrotactile feedback accompany the fracture of the virtual ice sheet underfoot. The two are derived from a simplified mechanical model. Fracture events are characterized via an event time t_i and energy loss E_i. Figure 8.5 illustrates the local continuum description and a simple mechanical analog used for synthesis. In this model, in the stuck state, the surface has stiffness $K = k_1 + k_2$ and is governed by:

$$F(t) = m\ddot{x} + b\dot{x} + K(x - x_0), \quad x_0 = k_2 \xi(t)/K \tag{8.3}$$

where $\xi(t)$ represents the net plastic displacement up to time t. A Mohr-Coulomb yield criterion determines slip onset: When the force F_ξ on the plastic unit exceeds a threshold F_0 (either a constant value or one sampled from a random process), a slip event is generated with energy loss E_i, representing the inelastic work of fracture growth. E_i is sampled from an exponential distribution $p(E) \propto E^{-\gamma}$ with a scale parameter γ that is, for many fracture processes, an approximate invariant of the material medium [309]. Slip displacements are rendered as transients given by a coupled model consisting of a nonlinear impulse coupled with a bank of modal

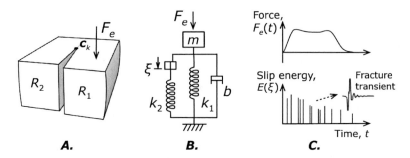

Fig. 8.5: A. Behavior at the crack front c_k is modeled using a simplified fracture mechanics treatment. A visco-elasto-plastic body undergoes shear sliding fracture. B. A simple mechanical analog. C. Each slip event is rendered as an impulsive transient.

oscillators with impulse response $s(t) = \sum_i a_i e^{-b_i t} \sin(2\pi f_i t)$, determined by amplitudes a_i, decay rates b_i, and frequencies f_i [22, 257]. A transient impulse at time t_0 is modeled phenomenologically as a nonlinear viscoelastic impact with effective force $f(t) \propto \Delta W$, simulated via the Hunt-Crossley impact model [129]

$$f(t) = kx(t)^\alpha - \lambda x(t)^\alpha \dot{x}(t). \tag{8.4}$$

$x(t)$ is the compression displacement and $\dot{x}(t)$ is the compression velocity. The impact has effective parameters governing stiffness k, dissipation λ, and contact shape α.

8.2.2.2 Visual animation and control

Brittle fracture in computer graphics is often animated by simulating the inelastic evolution of a distributed stress state [230, 213]. Here, we adopted a simplified simulation technique in order to fuse the local temporal crack-growth model given above with a heuristic for spatial crack pattern growth. The contact centroid x_c summarizes the local stress due to the load from a foot. A fracture pattern consists of a collection of crack fronts, defined by linear sequences of node positions, $c_0, c_1, \ldots, c_n$. Fronts originate at seed locations $p = c_0$. The fracture is rendered as line primitives $\ell_k = (c_k - c_{k-1})$ on the ice sheet (Figure 8.6). Seed locations p are determined by foot-floor contact. A crack event initiated by the audio-tactile process at time t_i with energy $E(t_i)$ results in the creation of a new seed or the growth of fractures from an existing one. In the former case, a new seed p is formed at the location of the dominant contact centroid x_c if no existing seed lies within distance Δp. The seed p is created with a random number N_c of latent crack fronts, $c_0^1, c_0^2, \ldots c_0^{N_c}$. We sample N_c uniformly in $2, 3, \ldots 6$, so that the cracks propagate outward from the initial contact position. A crack front propagates from a seed p nearest to x_c. With probability $1/N_c$ the jth crack front of p is extended. Its growth is determined by a propagation vector d_m^j such that $c_m^j = c_{m-1}^j + d_m^j$. We take $d_m^j = \alpha E \hat{n}_m^j$, where E is the crack energy, α

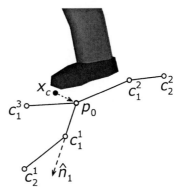

Fig. 8.6: A crack pattern,
modeled as a graph of lines
between nodes $\mathbf{c}_i$ extending
from the seed location p_0.

is a global growth rate parameter, and $\hat{\mathbf{n}}_m^j$ is the direction. Since we lack information
about the principal stress directions at the front, we propagate in a random direction
given by $\hat{\mathbf{n}}_m^j = \hat{\mathbf{n}}_{m-1}^j + \beta \hat{\mathbf{t}}$, where $\beta \sim N(\beta; 0, \sigma)$ is a Gaussian random variable and
$\hat{\mathbf{t}} = \hat{\mathbf{n}}^j \times \hat{\mathbf{u}}$, where $\mathbf{u}$ is the upward surface normal (i.e., $\mathbf{t}$ is a unit vector tangent to
$\hat{\mathbf{n}}^j$). The initial directions at $\mathbf{p}$ are spaced equally on the circle.

8.3 Walking on fluids

Water and other fluids have been largely ignored in the context of vibrotactile feedback. For VR simulations of real-world environments, the inability to include interaction with fluids is a significant limitation. The work described here represents
an initial effort to remedy this, motivated by our interest in supporting multimodal
VR simulations such as walking through puddles or splashing on the beach. Further
applications include improved training involving fluids, such as medical and phobia
simulators, and enhanced user experience in entertainment, such as when interacting
with water in immersive virtual worlds.

   To this end, we introduced the first physically based vibrotactile fluid rendering
model for solid-fluid interaction. Similar to other rendering approaches for virtual
materials, we profit from the fact that vibrotactile and acoustic phenomena share a
common physical source. Hence, we base the design of our vibrotactile model on
prior knowledge of fluid sound rendering. Since fluid sound is generated mainly
through bubble and air cavity resonance, we enhanced a fluid simulator with real-
time bubble creation and solid-fluid impact mechanisms, and can synthesize vibrotactile feedback from interaction and simulation events. Using this approach, we
explored the use of bubble-based vibrations to convey fluid interaction sensations to
users. We render the feedback for hand- or foot-based interaction, engendering rich
perceptual experiences of interacting with water features.

8.3.0.3 Acoustic models for fluid rendering

We aim at leveraging real-time fluid sound synthesis algorithms to generate the relevant vibrotactile feedback. Those techniques that are physically based rely on the oscillation of air bubbles trapped inside the fluid volume [255] to produce sound. The first bubble sound synthesis technique was proposed in Van den Doel's seminal work [79] where, based on Minnaert formula [187], he provides a simple algorithm to synthesize bubble sounds based on a few parameters. However, the synthesis was not coupled to a fluid simulation. This is achieved by Drioli et al. [81] through an ad-hoc model for the filling of a glass of water, based on the height of the fluid inside the glass and on collision events. Moss et al. [195] propose a simplified, physically inspired model for bubble creation, designed specifically for real-time applications. It uses the fluid surface curvature and velocity as parameters for bubble creation and a stochastic model for bubble sound synthesis based on Van den Doel's work [79]. However, the model is designed for a shallow water simulator, which greatly reduces interaction possibilities by allowing only surface waves, precluding splashes and object penetration.

Inspired by the physically based fluid sound synthesis work of Moss et al. [195], and utilizing a particle-based fluid model [192], we develop an efficient bubble generation technique and introduce a novel vibrotactile model. This enables rich body-fluid interactions with vibrotactile and multimodal cues.

8.3.0.4 Overview of the approach

When an object vibrates under an applied force, a pressure wave is generated at its surface, traveling to the subject's ears and mechanoreceptors. We motivate our vibrotactile approach, based on sound generation mechanisms, on the fact that acoustic and tactile feedback are both vibrations that share a common physical source.

By comparing film frames with the air-borne generated sound, Richardson [255] provides an explanation for the process of a projectile impacting and entering a fluid volume. The impact produces a "slap" and projects droplets, while the object penetration creates a cavity that is filled with air. The cavity is then sealed at the surface, creating an air bubble that vibrates due to pressure changes. Smaller bubbles can spawn from the fragmentation of the main cavity, as well as from the movement of the fluid-air interface, such as when the droplets return to the fluid volume.

Our vibrotactile model is therefore divided in three components, following the physical processes that generate sound during solid-fluid interaction [255, 98]: (1) the initial high frequency impact, (2) the small bubble harmonics, and (3) the main cavity oscillation. As a consequence, it is highly dependent on the efficient generation and simulation of air bubbles within the fluid. Hence, a real-time fluid simulator enhanced with bubble synthesis is required on the physical simulation side.

Figure 8.7 provides an overview of our approach. The physical simulator automatically detects the solid-fluid impacts and the creation of air bubbles caused by interaction between a solid (such as a foot, a hand or an object) and the fluid

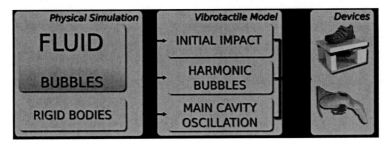

Fig. 8.7: Overview of our approach: the physical simulation computes the different parameters that are fed to the 3-step vibrotactile model, producing the signal sent to the various vibrotactile displays.

volume. For each of these events, it sends the corresponding message to the vibrotactile model, which synthesizes a vibrotactile signal according to the simulation parameters. The signal is then output through a specific vibrotactile device, such as an actuated tile for foot-fluid interaction, or a hand-held vibrator for hand-fluid interaction.

8.3.0.5 Fluid simulation with bubbles

The first building block of our approach is the fluid volume itself: we require a physically based real-time fluid simulation. Among existing fluid simulation techniques, the Smoothed-Particle Hydrodynamics (SPH) [192] model fulfills our requirements well, since the resulting fluid is unbounded, fast to compute and preserves small-scale details such as droplets. It is based on a set of particles discretizing the simulated media and conveying different physical properties, such as mass and viscosity. The motion of fluids is driven by the Navier-Stokes equations. Using the implementation of these equations in the SPH model [197], pressure and viscosity forces are computed at each time step. Rigid bodies are simulated as a constrained set of particles. For further details on SPH fluids, we refer the reader to [197].

As previously explained, in order to achieve vibrotactile interaction with fluids we need to simulate the bubbles inside the fluid. Since we only seek bubble creation events resulting in bubble sound synthesis, a bubble has a very short life span within our model, and can be seen more as an event than as the actual simulation of a pocket of air. Hence, we adopt and simplify an existing SPH bubble synthesis algorithm [198] to obtain an efficient bubble creation and deletion mechanism.

A bubble is spawned when a volume of fluid entraps a volume of air. In order to detect this phenomenon within the SPH simulation, we compute an implicit color field c^p as in the method of Muller et al. [198]. This color field estimates the amount of neighboring particles (fluid, rigid and bubble) around any position in space, while its gradient ∇c^p estimates in which direction the surrounding particles are mainly located. At each time step, we compute ∇c^p at each fluid particle position with:

$$\nabla c_i^p(\mathbf{x}_i) = \sum_j V_j \nabla W(\mathbf{x}_i - \mathbf{x}_j, h) \qquad (8.5)$$

A fluid particle i triggers a bubble creation if the following conditions are fulfilled:

1. The vertical component of ∇c_i^p is positive: the fluid particle has most of its surrounding particles above it, creating a pocket of air under it.
2. The magnitude of the velocity of the particle is above a threshold: still or slow moving fluid particles do not generate bubbles.

A bubble is destroyed when it is no longer entrapped by fluid or held by a rigid body. Since we only use bubbles for triggering events, a bubble is also destroyed if it is alone in the surrounding media. To this end, we compute another implicit color field, c^b, which only considers bubble particles, thus estimating the amount of neighboring bubbles surrounding a point in space. A bubble i is destroyed if one of the following conditions is fulfilled:

1. The vertical component of ∇c_i^p is negative: the bubble particle has most of its surrounding particles under it, and the air cannot be trapped anymore.
2. The color field c_i^b is null: the particle is alone inside the media.

8.3.0.6 Vibrotactile model

Our vibrotactile model receives the events from the physical simulation, and can synthesise a signal consisting of three components: the initial high frequency impact, the small bubble harmonics, and the main cavity oscillation. This is illustrated in Figure 8.8.

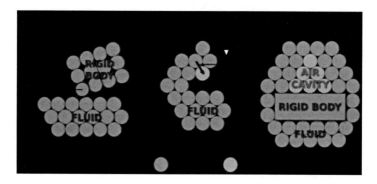

Fig. 8.8: The three components of our vibrotactile model.

Initial impact

During rigid body impacts on a fluid surface, Richardson [255] observed a damped high-frequency and low amplitude sound immediately after the impact, later explained as a guided acoustic shock [168]. To the best of our knowledge, no model provides the equations for air pressure oscillations due to a rigid body impact on a fluid surface. Previous work has been able to model the phenomenon to some extent, only for very simple shapes and specific penetration cases [128]. Nevertheless, the short duration of the impact does not justify a computationally expensive implementation. Hence, similar to previous work [81], we follow a physically inspired approach exploiting the short and burst-like nature of the vibration.

Synthesis

The impact signal is synthesized in a three step approach. A burst of white noise is first generated, spanning on the vibrotactile frequency range with a given base amplitude A. The signal is then fed to a simple envelope generator in order to modulate its amplitude. The signal rises exponentially during an attack time t_a, from nil to the original amplitude A, followed by an exponential decay of release time t_r, mimicking the creation and attenuation of the short and highly damped impact. Last, the modulated signal excites an elementary resonator. For this, a second-order resonant filter is used, creating a resonance peak around a central frequency w_0. The impact signal is therefore approximated as a resonating burst of white noise, with parameters to control its amplitude (A), duration (t_a, t_r) and central frequency (w_0).

Control

An impact event is triggered when the distance between a rigid body particle and a fluid particle is below the smoothing radius. Since only the particles at the surface of the rigid body have to be taken into account to avoid false triggers, a new implicit color field c^r is computed only considering rigid body particles: particles belonging to the lowest level sets of c^r belong to the surface. Richardson [255] observed that, in general, the intensity of the impact sound between a rigid body and a fluid is proportional to v^3, where v is the speed of the body at the moment of impact. Hence, after detecting an impact, we can synthesize an impact signal of amplitude A proportional to v^3. We use the same manually set duration and central frequency parameters for all impacts.

Harmonic bubbles

Small bubbles are generated by small pockets of air trapped under the water surface. Splashes and underwater cavity fragmentation are two causes for small bubble generation. By approximating all bubbles as spherical bubbles and relying on our SPH simulation enhanced with bubble generation, we can easily synthesize and control this component of the model.

Synthesis

Following van den Doel [79], we modeled the bubble as a damped harmonic oscillator. The pressure wave $p(t)$ of an oscillating spherical bubble is given by

$$p(t) = A_0 \sin(2\pi t f(t)) e^{-dt} \tag{8.6}$$

A_0 being the initial amplitude, $f(t)$ the resonance frequency and d a damping factor.

Minnaert's formula [187] approximates the resonance frequency f_0 of a spherical bubble in an infinite volume of water by $f_0 = 3/r$, where r is the bubble radius. In order to account for the rising in pitch due to the rising of the bubble towards the surface, Van den Doel [79] introduces a time dependent component in the expression of the resonant frequency: $f(t) = f_0(1 + \xi dt)$, with $\xi = 0.1$ found experimentally. Taking into account viscous, radiative and thermal damping, the damping factor d is set to $d = 0.13/r + 0.0072r^{-3/2}$. As for the initial amplitude A_0, previous work [174] suggests, after empirical observations, that $A_0 = \varepsilon r$ with $\varepsilon \in [0.01; 0.1]$ as a tunable initial excitation parameter. For a detailed explanation of the different hypotheses and equations, we refer the reader to [195] and [79].

Control

Our bubble vibration synthesis algorithm allows the generation of bubble sounds based on two input parameters: the bubble radius r and the initial excitation parameter ε. Using our SPH simulation, we couple the vibration synthesis with bubble creation events and automatically select the aforementioned parameters.

If we wanted to simulate the fluid and the bubbles at a scale where the particle radius matches the smallest bubble radius that generates a perceivable vibration (12 mm for a 250 Hz frequency), we would require around a million particles for a cubic meter of fluid. Achieving real-time performances with this number of particles is currently highly challenging for common hardware. Since we cannot directly link the particle radius to the resonating bubble radius, we transpose the physically inspired approach of Moss et al. [195] to determine the radius and excitation parameters from statistical distributions: power law distributions are used both for r and ε, within the ranges allowed by each parameter. When a bubble is created, values for r and ε are sampled, and sent to the signal synthesis algorithm.

Main cavity oscillation

The main cavity is a large bubble that produces a characteristic low-frequency bubble-like sound. By modeling the cavity as a single bubble with a large radius, we can rely on our harmonic bubble synthesis and control algorithms for this second component of our vibrotactile model.

Synthesis

As in the case of the harmonic bubble component, we use Equation 8.6 to synthesize the vibration produced by the oscillation of the main cavity. Since we will be using larger values for r, the resulting vibration will be of a much lower frequency, coherent with what we hear in real life. ε is set to 0.1 since no variability is desired.

Control

In order to detect the formation and collapsing of the main cavity during object penetration, we track the grouping of individual bubbles within our SPH simulation. Bubbles are spawned and stay alive when a cavity begins its closing and collapsing process, until they fill most of the cavity volume, as illustrated in Figure 8.8. At this point, there are bubbles within the cavity that are surrounded exclusively by other bubbles. These bubbles are detected when their color field c^b is above a threshold. If such a particle is detected, there is a potential cavity collapse.

Starting from the detected particle, we perform a search for neighboring bubbles to find the extent of the cavity. Bubble neighbors are added to the set of cavity bubbles, and the process is repeated on the new neighbors until no new neighbor is added. As the search is executed on the GPU, an iterative implementation is required, with one thread per new bubble neighbor, benefiting from our accelerated neighbor search algorithms. During our experiments, we required less than 5 search cycles to account for all the bubbles inside a cavity.

The total number N_b of cavity bubbles is proportional to the volume of the cavity. Since the cavity is modeled as a single large spherical bubble in the signal synthesis algorithm, its radius r can be deduced from the volume of the cavity. Hence, the number of cavity bubbles is mapped to the radius r of the spherical cavity, with user-defined minimal (r^{min}) and maximal (r^{max}) values: $[N_b^{min}, N_b^{max}] \rightarrow [r^{min}, r^{max}]$.

8.4 Multimodal rendering of fluids

Our fluid vibrotactile model is implemented in Puredata, while the SPH fluid and bubble simulation are implemented on GPU [62]. The communication between the SPH simulation and the acoustic model is handled through the Open Sound Control (OSC) protocol. Each time a bubble, cavity or impact event is detected in the fluid simulation, an OSC message is sent to the acoustic model with the corresponding parameters for sound synthesis. Since our vibrotactile model is built from sound generation mechanisms, we are able to produce acoustic feedback using the same model, by displaying the signal through a speaker and in the 12 Hz - 20 kHz range.

Kinesthetic feedback can also be rendered through a suitable haptic device, such as a multiple degrees-of-freedom force-feedback manipulator. The approach is described in previous work [62], using the same SPH fluid and rigid body simulation model.

We designed three scenarios representing three possible interaction conditions with multimodal feedback. See Figures 8.9 - 8.11. For the graphic rendering, we used a meshless screen-based technique optimized for high frequency rendering, described in previous work [62]. The scenarios were run on a Core 2 Extreme X7900 processor at 2.8 GHz, with 4 GB of RAM and an Nvidia Quadro FX 3600M GPU with 512 MB of memory.

Active foot-water interaction (shallow pool): This scene presents vibrotactile, acoustic and visual feedback (Figure 8.9). Our approach is particularly suited for foot-floor interaction, where the floor renders the vibrotactile feedback to the user's feet through appropriate vibrotactile transducers. Acoustic feedback can also be provided through speaker or headphones. We used a floor consisting of a square array of thirty-six 30.5×30.5 cm rigid vibrating tiles (see Chapters 2 and 3), rendering in the 20-750 Hz range. The virtual scene consisted of a virtual pool with a water depth of 20 cm filling the floor. The user's feet were modeled as parallelepiped rigid bodies and tracked through the floor pressure sensors. The user could walk about, splashing water as he stepped on the pool as seen in Figure 8.9. Performance: 15,000 particles (1% bubbles), 152Hz.

Fig. 8.9: Active foot-water interaction (shallow pool).

Passive foot-water interaction (beach shore): This scenario integrates vibro-tactile, acoustic and visual feedback (Figure 8.10). Using the same hardware setup as the previous scenario, we designed a tidal action scene in which the user stands still and experiences waves washing up on a sandy beach, as shown in Figure 8.10. Performance: 15,000 particles (6% bubbles), 147Hz.

Fig. 8.10: Passive foot-water interaction (beach shore).

Active hand-water interaction (water bucket): This scene incorporates vibro-tactile, kinesthetic, acoustic and visual feedback (Figure 8.11). The user can interact with fluids with his hands using a hand-held vibrotactile transducer and a 6DoF force feedback device. In this scenario, a small vibrator was attached to one of the user's hands. The hand was tracked by a motion capture system, and modeled in the virtual world as a parallelepiped rigid body. He could feel the water sensations by plunging his hand into a cubic volume of fluid, as seen in Figure 8.11. Figure 8.12 shows the vibrotactile signal generated during a plunging movement. Performance: 7,000 particles (6% bubbles), 240Hz.

Fig. 8.11: Active hand-water interaction (water bucket).

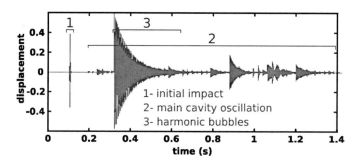

Fig. 8.12: Vibrotactile signal generated with our model during a plunging movement, with its three distinct components: (1) the initial impact, (2) the small bubble harmonics, and (3) the main cavity oscillation.

Chapter 9
Evaluation of multimodal ground cues

R. Nordahl, A. Lécuyer, S. Serafin, L. Turchet, S. Papetti, F. Fontana, and Y. Visell

Abstract This chapter presents an array of results on the perception of ground surfaces via multiple sensory modalities, with special attention to non visual perceptual cues, notably those arising from audition and haptics, as well as interactions between them. It also reviews approaches to combining synthetic multimodal cues, from vision, haptics, and audition, in order to realize virtual experiences of walking on simulated ground surfaces or other features.

9.1 Introduction

The multisensory perception of objects and surfaces that are felt or manipulated with the hands has been extensively studied in the literature, and this has, to some extent, informed the design of new generations of complex, multimodal human-computer interfaces that utilize touch, vision, and sound to access and interact with digital information or virtual worlds. As noted in the preceding chapters, substantially less research in either human perception or human-computer interaction has been devoted to interacting via the feet.

Multimodality is an increasingly common feature of interactive systems. Whilst most studies focus on the interaction between vision and audition or between vision and touch, interaction between touch and audition is also significant because both sources of sensory information possess high temporal resolution, and thus are produced by and evidence similar mechanical properties and interactions. Prior literature has investigated many aspects of audio-tactile cross-modal interactions in perception; see [164, 141, 269, 45]. Other, more applied, studies have investigated audio-tactile effects to enhance interaction with virtual worlds [248, 77, 76, 218, 274, 20].

An overview of the different studies considered in this chapter is given in Table 9.1, listing various experiments that have been developed on top of the technologies seen in the previous chapters.

Sensory Modality	Stimuli	Hypothesis
haptic + visual	camera motion + force feedback to the hands	vection illusion
auditory + visual	camera motion + loudspeaker listening	perception of bumps and holes
haptic (tactile) + auditory	vibrations underfoot + headphone listening	perception of bumps and holes
haptic (tactile) + auditory	vibrations underfoot + loudspeaker listening	path following
haptic (tactile) + haptic (kinesthetic)	variable compliance + vibrations underfoot	perception of ground stiffness
haptic (tactile) + auditory	vibrations underfoot + loudspeaker listening	tactile illusion underfoot
haptic (tactile) + auditory	vibrations underfoot + headphone listening	ground surface recognition
haptic (tactile) + auditory	vibrations underfoot + sound underfoot	effects on gait cycle

Table 9.1: Summary of information about the experiments described in the chapter.

The following sections contain results as well as references to more detailed descriptions of such experiments.

9.2 Salience of visual cues in ground perception

Vision is the best understood of the senses, and this is reflected in the preponderance of literature on self-motion perception, which has extensively investigated visual aspects. Here, we review a few novel experiments that together confirm that vision plays a leading role in framing our perception of ground surface properties. Where relevant and consistent with the visual feedback that is received, haptic and auditory cues can further contribute realism or other perceptual effects that would otherwise not be felt as strongly by perceivers.

9.2.1 Haptic Motion: Perception of self motion with force feedback and visual motion

"Haptic Motion" is a visuo-haptic paradigm for navigation in virtual worlds [217]. It allows users to feel their body being moved thanks to the application of force feedback to the hands in synergy to the projection of a scene reporting for self motion (see Figure 9.1).

We investigated the extent to which haptic forces felt through the hands can influence the perception of self motion, and how this influence compared with that of

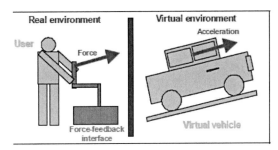

Fig. 9.1: Haptic Motion: force feedback corresponds to virtual acceleration.

visual stimulation alone. Our study involved both qualitative and quantitative measures, undertaken through two experiments. In the first, subjects were exposed to step-wise changes of *virtual acceleration*, rendered via for feedback to the hands and visual feedback, as in Figure 9.2. The visual acceleration that was supplied was

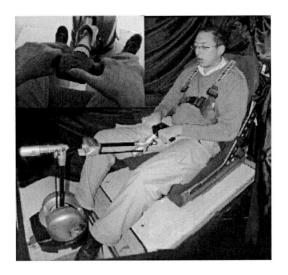

Fig. 9.2: Experimental apparatus used implementing Haptic Motion: force feedback to the user's hands is synchronized with visual feedback reporting for virtual self motion.

proportional to the haptic feedback (inertial force). Three experimental conditions were involved: haptic stimulation, visual stimulation and visuo-haptic stimulation. It was determined that the haptic force strongly influences the occurrence, onset and duration of a well-known effect in self-motion: the vection illusion.

To better understand how haptic information correlates visual information, in a second experiment we used different patterns of haptic force. It was observed that the haptic feedback evokes sensations of self motion in more complex 3D trajectories, and becomes important when subjects are exposed to force feedback that is proportional to the acceleration instead of speed.

Taken together, these results suggest that Haptic Motion could be used in various VR applications, to enhance sensation of self motion in VR and video games as well as in car driving simulators.

9.2.2 *Perception of bumps and holes with camera motion and footstep sounds*

Turchet et al. investigated the role of sound and vision in the recognition of different ground surface shapes, consisting of different configurations of bumps or holes along a virtual walking path [298]. Fifteen subjects participated in two within-subjects experiments. They were asked to interact with a desktop system displaying bumps, holes and flat surfaces by means of audio, visual and audio-visual cues. This display was similar to that used in purely visual experiments described in Section 6.4, allowing changes of the viewpoint in height (H), advance speed (V) and orientation (O), also simultaneously (HOV). Footstep sounds were synchronized with the vertical motion of the visual perspective rendered through the computer display, as determined by the virtual floor profile. The results of the first experiment show that participants were able to successfully identify the surface profiles in all conditions with very high success rates. The inclusion of auditory in fact did not produce higher percentages of recognition, which was already close to 100%.

In a second experiment, the dominance of vision was assessed, by presenting conflicting audio-visual stimuli. Results show that in presence of such conflicts audio is dominated by vision when H and O effects are presented. Conversely, vision is dominated by audio when V and HOV effects are presented. In particular the strongest dominance of the auditory modality was found when the visual stimuli were provided by means of the Velocity effect. Finally, a subjective questionnaire revealed a significant preference for the audio-visual stimuli compared to the unimodal condition.

9.3 Audio-haptic perception of virtual surface profiles

9.3.1 *Audio-haptic walking over bumps and holes*

Further to the experiments described in Section 9.2.2, simulations of auditory and haptic bumps or holes were presented to subjects as they were walking. In particular, it was investigated whether a variation of the IOI within and between footsteps, in both the auditory and haptic modality, affected the perception of surface inclination. This possibility is supported by the fact that people walking uphill tend to decelerate, while they accelerate when walking downhill.

While sitting on a chair subjects listened to footstep sounds through headphones, while feeling the corresponding vibrations through instrumented sandals. They were given a list of three different profile conditions (i.e. bump, hole, flat) presented as a forced alternative choice. The task consisted of recognizing a condition from each stimulus.

Forty-five participants were divided in three groups (n=15). These groups were composed respectively of 11 male and 4 female aged between 20 and 29 (mean =

23.6, std = 2.84), 11 men and 4 women, aged between 21 and 32 (mean = 24.86, std = 3.48) and 11 men and 4 women, aged between 20 and 28 (mean = 23.06, std = 2.40). All participants reported normal hearing conditions. They were naive with respect to the experimental setup and to the purpose of the experiment.

Results show that IOI variations between subsequent footsteps allow for successful recognition of bumps, holes, and flat surfaces especially thanks to the auditory modality. Furthermore, the inclusion of haptic cues significantly improves the recognition.

9.3.2 Walking on a virtual rope

An exploration on the role of auditory and haptic feedback in facilitating task performance was performed. The authors investigated whether these kinds of feedback facilitates the task of walking on a virtual rope, i.e. a particular case of path following. Subjects wearing instrumented sandals were blindfolded, and then asked to avoid falling from a virtual plank during an augmented walking task. Figure 9.3 shows a subject performing the experiment. Specifically, each subject was given the

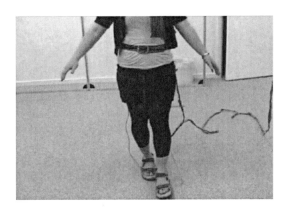

Fig. 9.3: Subject performing a walk on a virtual rope.

following instructions: "Imagine you are walking on a wooden plank. Your task is to walk from one side to the other. Walk slowly and pay attention to the feedback you receive. If your feet are outside of the plank you will fall." The auditory stimulation was designed in ways to simulate creaking wood when a user, whose position was detected by a motion capture system (Naturalpoint by Optitrack), was walking on top of the virtual plank.

The experiment was performed by 15 participants, 14 male and 1 female, aged between 22 and 28 (mean = 23.8, std = 1.97). All participants reported normal hearing. They were naive with respect to the experimental setup and to the purpose of the experiment. The results of the experiment did not provide clear indications on the role of the feedback to facilitate the task. Haptic cues appeared to be more salient, but differences with respect to the auditory feedback were not significant.

9.4 Nonvisual contribution to the perception of real and virtual ground material properties

The haptic perception of ground surface mechanical properties, such as softness or friction, or material types is essential in order to assure the stable regulation of dynamic posture and the control of locomotion in diverse environments. It is widely (often implicitly) assumed that kinesthetic (force-displacement) and visual perceptual cues dominate the sensorimotor control of locomotion over natural ground surfaces. However, a number of recent studies suggest that auditory and tactile cues acquired through the sole of the foot also contribute significantly to these perceptual processes.

9.4.1 Audio-haptic perception of virtual ground materials

Giordano et al. [110, 111] studied walkers' abilities to identify a variety of different walked-upon ground surfaces, comprising both solid materials (e.g., marble, wood) and granular media (e.g., gravel, sand) in different experimental conditions in which auditory, haptic, or audio-haptic information was available, and in a kinesthetic condition, where, during walking, tactile information was perturbed via vibromechanical noise to the sole of the foot. (Kinesthesia refers to the sense of movement and forces on the body.) Tactile masking was achieved using a novel shoe sole with integrated vibrotactile actuation (as described in Chapter 2).

The authors found haptic and audio-haptic discrimination abilities to be equally accurate, and determined that auditory and kinesthetic abilities to discriminate the ground surfaces studied are much less accurate. When walking on granular materials, which can shift underfoot, participants also appeared to focus preferentially on relatively inaccurate kinesthetic information when identifying the materials. The authors hypothesized that, although sub-optimal for the purpose of material identification, a focus on kinesthetic sensory channels indicates that attention was given preferentially to information that would most promptly signal postural instabilities.

9.4.2 Effect of plantar vibrotactile feedback on perceived ground stiffness

Visell et al. [311] investigated how the perception of ground surface compliance is altered by plantar vibration feedback. They conducted experiments in which 60 subjects walked in shoes over a rigid floor plate that provided supra- or near-threshold vibration feedback, and responded indicating how compliant it felt, either in subjective magnitude or via pairwise comparisons. In one experiment, the effect of plantar vibration feedback on ground compliance perception was measured through the use of a novel apparatus that allowed both the mechanical stiffness of a floor plate and vibration feedback presented through it to be manipulated (see Figure 9.4).

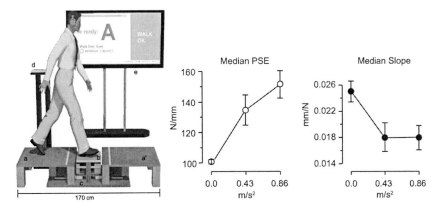

Fig. 9.4: Left: Vibrotactile floor interface from the experiment of Visell et al. [311]. Right: Point of subjective equality and psychometric curve slope for stiffness perception vs. vibration amplitude, based on fits to the experimental data.

Results showed that perceived compliance of the plate increased monotonically with vibration feedback intensity, and depended to a lesser extent on the temporal or frequency distribution of the feedback. When both plate stiffness (inverse compliance) and vibration amplitude were manipulated, the effect persisted, with both factors contributing to compliance perception. A significant influence of vibration was observed at low amplitudes (< 0.5 m/s^2) that were close to psychophysical detection thresholds for the stimuli. Taken together, the results of these experiments demonstrate that the perceived haptic compliance of a walking surface is increased in the presence of plantar cutaneous vibration feedback. The authors also found that an increased perception of compliance could be achieved with types of vibration feedback that differed in waveform, amplitude envelope, or the frequency distribution of their energy. None of the experiments involved training, and the effects observed did not require awareness that vibration feedback was being provided.

It was concluded that vibration felt during stepping on a rigid surface is combined with the mechanical stiffness of the surface in the haptic perception of compliance. In addition, the results show that the variation of vibration feedback alone is sufficient to elicit a percept of compliance. One hypothesis consistent with the observations is that plantar vibration feedback simulated the effect of increased displacement during stepping. This interpretation is also consistent with a basic mechanical description of the mechanics of material deformation underfoot during stepping. These findings show that vibrotactile sensory channels are highly salient to the perception of ground surface compliance, and suggest that correlations between vibro-tactile sensory information and motor activity may be of broader significance for the control of human locomotion than has been previously acknowledged.

9.4.3 Tactile illusion induced by low frequency auditory cues

Making use of the sandals described in Chapter 2, Papetti et al. investigated the influence of low-frequency auditory cues on the perception of underfoot vibration during a walking task. The results indicated that tactile perception is influenced by such cues. However, further experiments including more robust control conditions should be performed to add significance to such results. In this sense, they must be considered still preliminary.

Walking sounds from each shoe were routed to mini-speakers mounted on the shoes and to the vibrotactile transducers (haptuators) embedded in the respective sandals. Only their low frequency component was routed to four larger loudspeakers located at the corners of the experiment room, to avoid loss of footstep sound localization due to the auditory precedence effect. Finally, in order to enhance the sense of presence and the sound localization itself, environmental sounds of a forest (representing wind in the trees, birds singing and a river flowing) were superimposed to the auditory feedback from the loudspeakers.

Subjects wore the augmented sandals and walked at a regular pace along a pre-defined path. Halfway along the path, the intensity of the low frequency signal at the loudspeakers could be varied by ±6 dB or ±12 dB in the range [0,12] dB, or conversely left unchanged, with 0 dB corresponding to a loudspeaker loudness producing about 46 dB(A) in the room. Before the experiment, subjects were informed that this could occur, however they were not aware that only the audio feedback, and not the vibration, was altered. The experiment lasted about 45 minutes and consisted of twelve experimental configurations corresponding to all possible (both varied and unvaried) couples of low frequency levels. Each condition was repeated four times in balanced randomized order, for a total of 48 trials. After each trial, subjects had to write down whether they had felt any change in the vibrotactile feedback under their feet (answer: yes/no). For each participant, the percentages of "yes" responses were calculated for the twelve experimental conditions. The difference from random percentage (50%) was tested using one-proportion (two-tailed) z tests, and we used two-proportion (two-tailed) z-tests in order to check the differences between the experimental conditions.

The experimenters considered the percentages of "yes" responses for each couple of stimuli. The results are presented in Figure 9.5. As expected, the largest low frequency variation (amounting to ±12 dB) corresponded to the strongest illusion. On the other hand, it was found that couples introducing a variation of ±6 dB resulted in considerably different effects, indicating that the corresponding intensity changes are possibly too small to firmly overcome the existing thresholds of illusory tactile detection underfoot. In absence of a further experiment including a more robust con-

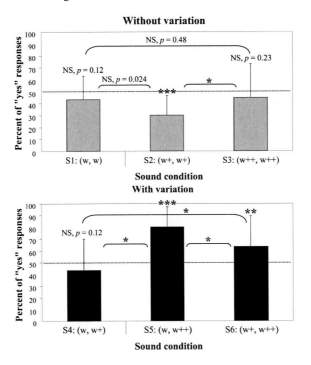

Fig. 9.5: (Above) Mean percentage of "yes" responses (bars represent std) for the unvaried couples as a function of the sound configurations S1: (w,w), S2: (w+,w+), S3: (w++,w++), with w=0 dB, w+=±6 dB, w++=±12 dB. The difference from random (line at 50%) was tested using one-proportion (two-tailed) z-tests. The differences between the three sound conditions were tested with two-proportion z-tests (two-tailed and Bonferroni-adjusted alpha-level with p = 0.05/3 = 0.0167). (Below) Mean percentage of "yes" responses (bars represent std) for the varied couples as a function of the sound configurations S4: (w,w+), S5: (w,w++), S6: (w+,w++), with w=0 dB, w+=±6 dB, w++=±12 dB. The difference from random (line at 50%) was tested using one-proportion (two-tailed) z-tests. The differences between the three sound conditions were tested with two-proportion z-tests (two-tailed and Bonferroni-adjusted alpha-level with p = 0.05/3 = 0.0167). Legend: *:$p < 0.05$, **: $p < 0.01$, ***:$p < 0.001$, NS: not significant.

trol test, these results suggest that a cross-modal effect is present which manifests itself as an audio-tactile illusion, where audio low frequency influences vibrotactile perception.

9.4.4 Audio-haptic identification of ground surfaces

This section overviews experiments whose goal was to investigate the subjective ability to recognize auditory and haptic ground surface simulations. All these experiments were carried out in an acoustically isolated laboratory sized approximately

18 square meters, allowing subjects to walk with the instrumented sandals (Chapter 2) while wearing a pair of Sennheiser HD 650 headphones.

In the auditory conditions, the haptic actuators of the sandals were not used. The pressure sensors inside the sandals were used to drive the audio synthesis engine. In the haptic conditions, participants wore earplugs and sound protection headsets instead of headphones, to minimize any external sound interference.

Offline experiments were conducted by having subjects sit and experience feedback provided to the shoes or to the ears. Conversely, online experiments were conducted by allowing subjects to walk across the laboratory, hence enabling the interactive features of the sandals (see Figure 9.6).

Fig. 9.6: A person wearing the sandals enhanced with pressure sensors and actuators.

9.4.4.1 Offline audio-haptic identification of virtual grounds

The goal of this experiment was to assess whether subjects are able to recognize virtual ground surfaces from offline audio or haptic cues. More details on the experiment are described in [209].

All participants were asked to wear a pair of instrumented sandals and headphones, and then to sit on a chair. The task consisted of recognizing a ground surface from passively felt audio-haptic stimuli. They were given a list of sixteen materials: wood, creaking wood, underbrush, snow, frozen snow, beach sand, gravel, metal, high grass, dry leaves, concrete, dirt, puddles, water, carpet and 'I don't know'. Each material was presented twice in a random sequence. Participants had to match every stimulus to an item in the list, furthermore to rated the realism and quality of the simulations. to debrief.

Forty five volunteers (students and faculty members of the Engineering college in Copenhagen; 31 male and 14 female; average age = 24.5, std = 4.6) were randomly assigned to one of these three groups: audio, haptic, audio-haptic. None reported hearing problems or other sensory impairments. The results indicated that haptic cues alone enabled poor discrimination of ground surfaces. Though, solid surfaces were not confused with aggregate, and vice-versa. Furthermore, the combination of auditory and haptic cues did not result in better recognition performance.

9.4.4.2 Online audio-haptic identification of virtual grounds

The goal of this experiment was to assess whether subjects were able to recognize virtual ground surfaces from online audio and haptic cues.

Thirty participants were divided in three groups (n = 10) to perform a between-subjects experiment. The three groups were composed respectively of 7 male and 3 female aged between 20 and 35 (mean = 24.6, std = 4.67), 9 male and 1 female aged between 20 and 31 (mean = 23.4, std = 3.23), and 7 male and 3 female aged between 21 and 25 (mean = 22.7, std = 1.07). All participants reported normal hearing.

Groups 1 and 3 wore the instrumented sandals and headphones. Group 2 wore the same sandals, along with earplugs and sound protection headsets. All groups then performed a walking task across the laboratory.

Eight stimuli were presented twice in randomized order. The stimuli consisted of audio and haptic simulations of footstep sounds on the following surfaces: beach sand, gravel, deep snow, forest underbrush, dry leaves, wood, creaking wood, metal. Participants were given a list of ground surfaces in form of a non-forced alternate choice, included also materials which were not present in the set of stimuli.

Subjects simultaneously perceived footsteps sounds and/or vibrations during spontaneous walking tasks. The results confirm that recognition was more success-ful compared to the previous experiment. As in the offline case, the combination of auditory and haptic stimuli did not significantly enhance the recognition. More details on the experiment are described in [267].

9.4.4.3 Audio-haptic matching of ground categories

A between-subjects experiment was conducted, whose goal was to investigate pos-sible dominance of the audio or haptic modality during an augmented walking task. Subjects were asked to recognize surface material sounds and vibrations during the task. Both coherent and incoherent stimuli were presented in form of audio-haptic couples of surface materials. Incoherent couples contained materials belonging to different categories: if the auditory feedback reported for a solid surface, the simul-taneous haptic feedback was of an aggregate surface and vice-versa. The hypoth-esis was that the audio modality dominates over the haptic one. Another was that the recognition would have slightly improved using coherent rather than incoherent stimuli.

As previously described, participants were asked to wear a pair of instrumented sandals and headphones, then to walk across the laboratory. During walking they simultaneously perceived footstep sounds and vibrations. The task consisted of recognizing the surfaces they were exposed to. As opposed to the previous experiments, participants were not provided with a forced list of possible choices.

Participants were exposed to 12 trials consisting of 4 coherent stimuli and 8 incoherent stimuli. The 12 audio-haptic stimuli were presented once in randomized order. The modeled surfaces were 4 (2 solid and 2 aggregate): wood, metal, snow and gravel. All possible material incoherences existing between the two categories were accounted for by the 8 stimuli, for both modalities.

Ten participants, 7 male and 3 female, aged between 20 and 38 (mean = 25.81, std = 5.77), were involved in the experiment. All participants reported normal hearing conditions and all of them were naive with respect to the experimental setup and to the purpose of the experiment.

Results show that the auditory modality dominates over the haptic one: in both coherent and incoherent conditions, subjects tend to classify the floor surface category by listening. Furthermore, coherent audio-haptic presentations of a surface material do not result in significantly improved subjective performance. More details on the experiment are described in [301].

9.5 Effects of ecological auditory and vibrotactile underfoot feedback on human gait: a preliminary investigation

A pilot experiment was carried out [226] in which individual IOIs were measured while subjects were asked to walk along a predefined path while wearing the audio-tactile instrumented sandals of Chapter 2. The experimental hypothesis was that human gait can be influenced by providing ecological audio-tactile feedback through the feet. Virtual snow and mud were presented based on the physics-based models of Chapter 7, along with one neutral (i.e. control) condition presenting no artificial feedback. Eight subjects, seven males and one female, participated in the experiment. Their average age was 22.3 years. Six subjects were right-handed and also considered their right foot as dominant. One subject was left-handed and one ambidextrous, also with regard to the use of his feet. None of them reported locomotion disorders.

The participants were asked to wear the instrumented sandals and included backpack, and walked along an eight-shaped trajectory in the experiment room. They were informed that a change in the multimodal feedback could occur at each trial, and no other instructions were given. In order to avoid biases due to the room configuration, half of the participants started walking from one of the shorter sides of the rectangular room and the other half from the other side. The experiment considered 8 trials for each condition, resulting in 24 trials in balanced randomized order for each experimental session, which lasted about 15 minutes.

The following IOIs were analyzed starting from measured ground reaction force thresholds: within-foot heel-to-toe intervals for both the left and right foot; heel-to-heel intervals; between-feet heel-to-toe intervals. A one-way repeated measures ANOVA (RM-ANOVA) was first performed on the data recorded in the neutral condition, in order to verify whether the subjects walked with a regular pace when no stimuli were present. The different trials in this condition were considered as repeated measures for each subject, and the mean IOIs in each trial were used for the analysis. The obtained p-values are very high, meaning that the subjects walked with a regular gait in the neutral condition. In particular, the heel-to-toe IOI for the right foot appeared to be extremely regular and was chosen as reference.

The same IOI under conditions of virtual snow and mud was analyzed using a RM-ANOVA, again considering the mean IOI over all trials under each condition. Results show that the subjects' gait was slightly affected by the virtual feedback: the IOIs in fact are more irregular compared to the neutral condition.

Indeed, the results are close to statistical significance. However, the relatively high p-values indicate that the effects need further investigation. In particular, the experiment should be repeated with a larger number of subjects. In a broader perspective, the lack of a clear statistical significance in the obtained results may have an alternative interpretation. In fact the ability to provide salient non-visual cues underfoot, which do not significantly alter one's walking style, may enable the design of foot interfaces supplying informative, meanwhile non-intrusive messages for guiding users across spaces otherwise difficult to navigate.

9.6 Conclusions

This chapter presented an overview of multimodal experiments performed in the context of foot-floor interactions. The experiments were performed with the goal of evaluating different simulation technologies, while at the same time achieving a better understanding of the role of the sensory modalities in the discrimination of surface textures and ground properties, also in the context of cross-modal illusions.

In the limits of the (often debatable) statistical significances, some general rules can be learned from these experiments:

- subjects can recognize simulated surfaces using both auditory and haptic cues;
- the combination of auditory and haptic information does not significantly enhance the recognition;
- subjects are able to recognize simulated surface profiles that are reproduced visually, auditorily and haptically;
- auditory and haptic feedback slightly modifies a subject's gait, although not significantly.

Using auditory and haptic feedback also allows to recreate some illusions, such as a sensation of stronger tactile cues when only auditory feedback is boosted. Moreover, auditory and haptic feedback can be used to signal to subjects to walk on a given path, such as a straight line.

Taken together, these experiments provide some evidence of the importance of floor feedback in simulated environments, furthermore they call for more research on a topic which has been rather unexplored in the virtual reality community.

Conclusion

F. Fontana and Y. Visell

Although the research reviewed in the preceding chapters provides some counterexamples, the range of human sensory abilities, and, in particular, the excellent somatosensory capacities of the human feet, have been under-utilized in human-computer interfaces, and the concept of using shoes and floors as rich, multimodal interfaces has not reached its full potential in contemporary research and development. This book is among the first systematic attempts to collect a diverse variety of results in the broad field of human-computer interaction research with an aim to increase awareness of the potential for the foot and locomotion to enhance and play a more active role as part of the human interface with information technologies.

Although much of the work presented here is preliminary, being based on technologies or methodologies that are still under development, the editors hope that some key ideas of these researchers, which were not fully formed as recently as a few years ago, have become more accessible through the publication of this book. As was noted in the introduction, some of the unresolved issues reflected in the relatively raw presentation of some of the material reflects a decision to include very recent research results that have left key questions open, to be resolved in future research. To the extent that this limitation has been detrimental to the quality of this volume, it is hoped that this shortcoming is partly compensated by the timeliness of its publication and by its public accessibility.

At the same time, experimental results reviewed in this volume have already directly informed the invention of new kinds of multimodal displays for realizing computationally augmented experiences centered on locomotion. In particular, the identification of remarkable cross-modal perceptual illusions, such as those relating perceived ground shape during walking in a virtual environment to camera viewpoint distortions, or linking perceived ground compliance to the felt intensity of vibration feedback delivered to the soles of the feet, suggest that novel displays exploiting fundamental properties of the human perceptual apparatus may enable virtual walking displays that are low in cost and complexity, or that utilize existing display devices.

The opportunity to design ecological feedback in response to walking with a focus on non-visual modalities suggests a complementary point of view to prevailing trends in mobile information appliances. Floor-based interfaces could be used to provide orientation cues to pedestrians, and, perhaps, simultaneously harvest energy exerted by human walkers. For example, as recently as October 2011, it was reported in the IEEE Spectrum magazine that researchers had developed electric shoe insoles that made it possible to recharge the batteries in a pair of shoes through walking. In parallel, novel mobile configurations for supplying simple haptic feedback through the shoes are growing, and interest in them may be further enhanced if they are considered to be useful displays that can exchange location-based information via a user's smartphone.

Floor- and shoe-based interactive devices have proliferated in the worlds of entertainment, gaming, rehabilitation and sports training, via in the form of embodiments such as visually interactive surfaces, instrumented balance boards, or electronic shoes that enable functions ranging from the measurement of kinematic parameters during locomotion to the control of music through dancing. While these developments may seem to be less directly linked to the results presented here, and somewhat outside the focus of this volume, the book may serve to further stimulate innovation in these areas.

From a longer term perspective, the interactive augmentation of flooring and footwear with multimodal ecological cues simulating those encountered in real walking experiences (e.g., variations in the feel of ground materials underfoot, or other virtual changes in the terrain) may give rise to robust interaction paradigms and devices that could become progressively integrated in everyday, real-world experiences. This could, in turn, fundamentally impact the ways in which we acquire information from our surroundings in significant ways. Although this vision remains remote from current possibility, some strong points in its favor include: the ecological unity and intuitive nature of the ways that information can be delivered; its potential exportability to diverse cultures, language, and social systems; the potential benefits of digital information accessibility to persons with sensory or even cognitive deficits of different types; and the potentially ubiquitous, calm, and non-distracting nature of the interface, as in the case of tactile interfaces that are only felt by a single user.

Another arena in which interactive shoe or floor interfaces may bring advantages is for the design of urban spaces, whose navigation is frequently non-obvious, despite long-standing traditions of visual and auditory signaling as well as the design of passive haptic markers underfoot to demarcate everything from recreational trails to pedestrian crosswalks, railway platforms, and subway entrances, frequently with the aim, noted above, of making the navigation of public and private spaces more accessible to persons with sensory disabilities.

Moreover the rendering and display of specific floor cues may have clinical applications toward assisting the recovery of stroke patients or other disorders affecting sensorimotor function during locomotion. Indeed, there is growing interest in the design of new sensory feedback techniques and rehabilitation paradigms to enhance the ability of the nervous system to repair itself following injury. The availability of new devices that can enable to controlled presentation of novel stimuli for the feet, and reinforce the ecological content in somatosensory cues conveyed to otherwise impaired bodily systems may encourage further clinical investigations aimed at the creation of potentially individually adapted rehabilitation therapies based on selective stimulation of the feet. Finally, there may also be potential benefits for cases of permanent sensory impairments, either through enhancements that assist visually- or hearing-impaired individuals in important activities, such as travel or negotiation of built environments, or in settings, such as those potentially encountered in the space sciences, where humans must operate under dangerous conditions without their normal sensory faculties available to them. Similar interventions could be applicable toward assisting activities in other sensory-deprived settings, such as undersea labor or exploration, or in diverse areas of robotic teleoperation.

It is hoped that the efforts made by the authors and editors of this book may contribute toward realizing of these prospective visions in the future.

References

1. A. Abu-El-Quran, R. A. Goubran, and A. D. C. Chan. Security monitoring using microphone arrays and audio classification. *IEEE Trans. on Instrumentation and Measurement*, 55(4), 2006.
2. M. Addlesee, A. Jones, F. Livesey, and F. Samaria. The orl active floor. *IEEE Personal Communications*, 4(5):35–51, 1997.
3. M. Addlesee, A. H. Jones, F. Livesey, and F. S. Samaria. The ORL active floor. *IEEE Personal Communications*, 4:35–41, 1997.
4. F. Adjémian and P. Evesque. Different regimes of stick-slip in granular matter: from quasi-periodicity to randomness. *EGS XXVII General Assembly, Nice, 21-26 April 2002*, 2002.
5. J. M. Adrien. The missing link: Modal synthesis. pages 269–297, 1991.
6. P. Agrawal, I. Rauschert, K. Inochanon, L. Bolelli, S. Fuhrmann, I. Brewer, G. Cai, A. MacEachren, and R. Sharma. Multimodal interface platform for geographical information systems (geomip) in crisis management. In *ICMI '04: Proceedings of the 6th international conference on Multimodal interfaces*, pages 339–340, New York, NY, USA, 2004. ACM.
7. H. J. Ailisto, M. Lindholm, J. Mantyjarvi, E. Vildjiounaite, and S.-M. Makela. Identifying people from gait pattern with accelerometers. In *Proc. SPIE – Biometric Technology for Human Identification II*, 2005.
8. M. Alava, P. Nukala, and S. Zapperi. Statistical models of fracture. *Advances in Physics*, 55(3):349–476, 2006.
9. P. Albinsson and S. Zhai. High precision touch screen interaction. In *Proceedings of the SIGCHI conference on Human factors in computing systems*, pages 105–112. ACM, 2003.
10. K. Aminian and B. Najafi. Capturing human motion using body-fixed sensors: outdoor measurement and clinical applications. *Journal of Visualization and Computer Animation*, 15(2):79–94, 2004.
11. R. Annies, E. Martinez Hernandez, K. Adiloglu, H. Purwins, and K. Obermayer. Classification schemes for step sounds based on gammatone-filters. 2007.
12. G. Arechavaleta, J.-P. Laumond, H. Hicheur, and A. Berthoz. The nonholonomic nature of human locomotion: a modeling study. In *Proceedings of IEEE International Conference on Biomedical Robotics and Biomechatronics*, pages 158–163, 2006.
13. S. Arulampalam, S. Maskell, and N. Gordon. A tutorial on particle filters for online nonlinear/non-gaussian bayesian tracking. *IEEE Transactions on Signal Processing*, 50:174–188, 2002.
14. D. Ashbrook and T. Starner. Learning significant locations and predicting user movement with gps. In *International Symposium on Wearable Computing*, Seattle, WA, October 2002.
15. D. Ashbrook and T. Starner. Using gps to learn significant locations and predict movement across multiple users. *Personal Ubiquitous Comput.*, 7(5):275–286, 2003.
16. K. J. Astrom. Optimal control of Markov decision processes with incomplete state estimation. *J. Math. Anal. Applic.*, 10:174–205, 1965.

17. J.-J. Aucouturier, B. Defreville, and F. Pachet. The bag-of-frame approach to audio pattern recognition: A sufficient model for urban soundscapes but not for polyphonic music. *Journal of the Acoustical Society of America*, 2007.

18. J.-J. Aucouturier, F. Pachet, P. Roy, and A. Beurivé. Signal + context = better classification. In *Proceedings of ISMIR 07*, pages 425–430, 2007.

19. T. Augsten, K. Kaefer, R. Meusel, C. Fetzer, D. Kanitz, T. Stoff, T. Becker, C. Holz, and P. Baudisch. Multitoe: High-precision interaction with back-projected floors based on high-resolution multi-touch input. In *Proceedings of the 23nd annual ACM symposium on User interface software and technology*, pages 209–218. ACM, 2010.

20. F. Avanzini and P. Crosato. Integrating physically based sound models in a multimodal rendering architecture: Research articles. *Comput. Animat. Virtual Worlds*, 17(3-4):411–419, 2006.

21. F. Avanzini and D. Rocchesso. Modeling collision sounds: non-linear contact force. In *Proc. Conf. on Digital Audio Effects (DAFX-01)*, pages 61–66, Limerick, December 2001.

22. F. Avanzini and D. Rocchesso. Modeling Collision Sounds: Non-linear Contact Force. *Proc. COST-G6 Conf. Digital Audio Effects (DAFX-01)*, pages 61–66, 2001.

23. F. Avanzini, S. Serafin, and D. Rocchesso. Interactive simulation of rigid body interaction with friction-induced sound generation. *Speech and Audio Processing, IEEE Transactions on*, 13(5):1073–1081, 2005.

24. E. Ayyappa. Normal human locomotion, part 2: Motion, ground reaction force. *J. Prosthetics and Orthotics*, 9(2), 1997.

25. A. Baldassarri, F. Dalton, A. Petri, S. Zapperi, G. Pontuale, and L. Pietronero. Brownian forces in sheared granular matter. *Physical Review Letters*, 96(11):118002, 2006.

26. L. Bao. Physical activity recognition from acceleration data under semi-naturalistic conditions. Technical report, Master's thesis, MIT, 2003.

27. L. Bao and S. Intille. Activity recognition from user-annotated acceleration data. *Pervasive 2004*, pages 1–17, April 2004.

28. S. Barrass and M. Adcock. Interactive granular synthesis of haptic contact sounds. In *AES 22nd International Conference on Virtual, Synthetic and Entertainment Audio*, 2002.

29. C. Basdogan and M. A. Srinivasan. *Handbook of Virtual Environments: Design, Implementation, and Applications*. CRC Press, 1 edition, Jan. 2002.

30. S. Basu, S. J. Schwartz, and A. Pentland. Wearable phased arrays for sound localization and enhancement. In *ISWC*, page 103, 2000.

31. L. E. Baum, T. Petrie, G. Soules, and N. Weiss. A maximization technique occurring in the statistical analysis of probabilistic functions of markov chains. *Ann. Math. Statist.*, 41(1):164–171, 1970.

32. R. K. Begg, M. Palaniswami, and B. Owen. Support vector machines for automated gait classification. *IEEE Transactions on Biomedical Engineering*, 52(5), 2005.

33. F. Behrendt, L. Gaye, and A. Tanaka, editors. *Proc. of the 4th Intl. Mobile Music Workshop*, Amsterdam, NL, May 2007.

34. H. Benko, A. Wilson, and P. Baudisch. Precise selection techniques for multi-touch screens. In *Proceedings of the SIGCHI conference on Human Factors in computing systems*, pages 1263–1272. ACM, 2006.

35. B. Bhanu and X. T. Zou. Moving humans detection based on multi-modal sensor fusion. In *Object Tracking and Classification Beyond the Visible Spectrum*, page 136, 2004.

36. A. Bicchi, J. Salisbury, and D. Brock. Contact sensing from force measurements. *The Int J of Robotics Research*, 12(3):249, 1993.

37. C. M. Bishop. *Pattern Recognition and Machine Learning*. Springer, August 2007.

38. A. Bissacco, A. Chiuso, Y. Ma, and S. Soatto. Recognition of human gaits. *Computer Vision and Pattern Recognition, IEEE Computer Society Conference on*, 2:52, 2001.

39. A. Bissacco and S. Soatto. Modeling and learning contact dynamics in human motion. In *Proc. of Computer Vision and Pattern Recognition (CVPR)*, 2005.

40. R. E. Bland. Acoustic and seismic signal processing for footsetp detection. Master's thesis, Massachusetts Institute of Technology. Dept. of Electrical Engineering and Computer Science., 2006.

41. B. Blesser. An interdisciplinary synthesis of reverberation viewpoints. *J. of the Audio Engineering Society*, 49(10), 2001.

42. G. Borin, G. De Poli, and D. Rocchesso. Elimination of Delay-Free Loops in Discrete-Time Models of Nonlinear Acoustic Systems. *IEEE Trans. on Speech and Audio Processing*, 8(5):597–605, September 2000.

43. D. A. Bowman, E. Kruijff, J. J. LaViola, and I. Poupyrev. *3D User Interfaces: Theory and Practice*. Addison-Wesley Professional, 1 edition, 2004.

44. G. R. Bradski. Computer vision face tracking for use in a perceptual user interface. *Intel Technology Journal*, 2(2):12–21, 1998.

45. J.-P. Bresciani, M. O. Ernst, K. Drewing, G. Bouyer, V. Maury, and A. Kheddar. Feeling what you hear: Auditory signals can modulate tactile tap perception. *Experimental Brain Research*, 162:172–180, 2005.

46. R. Bresin, A. de Witt, S. Papetti, M. Civolani, and F. Fontana. Expressive sonification of footstep sounds. In R. Bresin, T. Hermann, and A. Hunt, editors, *Proc. of the Interaction Sonification workshop (ISon) 2010*, KTH, Stockholm, Sweden, Apr. 7 2010.

47. R. Bresin, S. Delle Monache, F. Fontana, S. Papetti, P. Polotti, and Y. Visell. Auditory feedback from continuous control of crumpling sound synthesis. In *CHI 2008 Workshop on Sonic Interaction Design*, Florence, Italy, Apr. 5-10 2008. ACM.

48. M. Brubaker, L. Sigal, and D. Fleet. Estimating Contact Dynamics. In *IEEE International Conference on Computer Vision (ICCV)*, 2009.

49. G. Bruder, F. Steinicke, and K. H. Hinrichs. Arch-Explore: a natural user interface for immersive architectural walkthroughs. In *IEEE Symposium on 3D User Interfaces, 2009. 3DUI 2009*, pages 75–82. IEEE, 2009.

50. M. Büchler, S. Allegro, S. Launer, and N. Dillier. Sound classification in hearing aids inspired by auditory scene analysis. 2005.

51. G. Burdea and M. Akay. *Force and touch feedback for virtual reality*. Wiley New York, 1996.

52. M. Casey. General sound classification and similarity in mpeg-7. *Organised Sound*, 6(2):153–164, 2001.

53. M. Casey. *Sound Classification and Similarity, in: Introduction to MPEG-7, Multimedia Content Description Interface*. John Wiley and Sons, Ltd., Jun 2002.

54. G. Casiez, D. Vogel, Q. Pan, and C. Chaillou. RubberEdge: reducing clutching by combining position and rate control with elastic feedback. In *Proceedings of the ACM symposium on User interface software and technology*, pages 129–138, Newport, Rhode Island, USA, 2007. ACM.

55. T. Chau. A review of analytical techniques for gait data. part 1: fuzzy, statistical and fractal methods. *Gait and Posture*, 13(1), 2001.

56. T. Chau. A review of analytical techniques for gait data. part 2: neural network and wavelet methods. *Gait and Posture*, 13(2), 2001.

57. J. Chen, A. H. Kam, J. Zhang, N. Liu, and L. Shue. Bathroom activity monitoring based on sound. *Pervasive Computing*, pages 47–61, 2005.

58. L. Cheng and S. Hailes. On-body wireless inertial sensing for foot control applications. In *IEEE 19th International Symposium on Personal, Indoor and Mobile Radio Communications 2008 (PIMRC 2008)*, 2008.

59. M. Chion. *Audio-Vision*. Columbia Univ. Press, New York, USA, 1994.

60. R. Christensen, J. Hollerbach, Y. Xu, and S. Meek. Inertial-force feedback for the treadport locomotion interface. *Presence: Teleoperators & Virtual Environments*, 9(1):1–14, 2000.

61. P. Chueng and P. Marsden. Designing Auditory Spaces to Support Sense of Place: The Role of Expectation. *CSCW Workshop: The Role of Place in Shaping Virtual Community*, 2002.

62. G. Cirio, M. Marchal, S. Hillaire, and A. Lecuyer. Six Degrees-of-Freedom haptic interaction with fluids. *IEEE Transactions on Visualization and Computer Graphics*, PP(99):1–1, Nov. 2011.

63. M. Civolani, F. Fontana, and S. Papetti. Efficient acquisition of force data in interactive shoe designs. In *Proc. 5th Int. Haptic and Auditory Interaction Design Workshop*, pages 129–138, Sept. 2010.

64. B. P. Clarkson and A. Pentland. Extracting context from environmental audio. In *ISWC*, pages 154–155, 1998.

65. P. R. Cook. Modeling Bill's gait: Analysis and parametric synthesis of walking sounds. In *Proc. Audio Engineering Society 22 Conference on Virtual, Synthetic and Entertainment Audio*, Espoo, Finland, July 2002. AES.

66. G. Courtine and M. Schieppati. Human walking along a curved path. i. body trajectory, segment orientation and the effect of vision. *Eu. J. of Neuroscience*, 18(1):177–190, 2003.

67. D. H. Cress. Terrain considerations and data base development for the design and testing of devices to detect intruder-induced ground motion. Technical Report M-78-1, U.S. Army Engineer Waterways Experiment Station, Vicksburg, Miss., 1978.

68. A. Crevoisier and P. Polotti. Tangible acoustic interfaces and their applications for the design of new musical instruments. In *NIME '05: Proceedings of the 2005 conference on New interfaces for musical expression*, pages 97–100, Singapore, Singapore, 2005. National University of Singapore.

69. A. Crossan, J. Williamson, and R. Murray-Smith. Haptic granular synthesis: Targeting, visualisation and texturing. In *Proc. Intl Symposium on Non-visual & Multimodal Visualization*, pages 527–532, 2004.

70. B. Dalton and M. Bove. Audio-based self-localization for ubiquitous sensor networks. In *118th Audio Engineering Society Convention*, 2005.

71. F. Dalton, F. Farrelly, A. Petri, L. Pietronero, L. Pitolli, and G. Pontuale. Shear stress fluctuations in the granular liquid and solid phases. *Physical Review Letters*, 95(13):138001, 2005.

72. R. Darken, W. Cockayne, and D. Carmein. The omni-directional treadmill: a locomotion device for virtual worlds. In *Proceedings of the 10th annual ACM symposium on User interface software and technology*, pages 213–221. ACM, 1997.

73. A. de Groot, R. Decker, and K. Reed. Gait enhancing mobile shoe (gems) for rehabilitation. In *Proceedings of World Haptics*, pages 190–195, 2009.

74. S. Delle Monache, P. Polotti, and D. Rocchesso. A toolkit for explorations in sonic interaction design. In *Proc. of the 5th Audio Mostly Conf.*, AM '10, pages 1:1–1:7, Pitea, Sweden, 2010. ACM.

75. A. P. Dempster, N. M. Laird, and D. B. Rubin. Maximum likelihood from incomplete data via the EM algorithm. *Proceedings of the Royal Statistical Society*, B-39:1–38, 1977.

76. D. DiFilippo and D. K. Pai. Contact interaction with integrated audio and haptics. In *Proceedings of the International Conference on Auditory Display, ICAD*, 2000.

77. D. E. DiFranco, G. L. Beauregard, and M. A. Srinivasan. The effect of auditory cues on the haptic perception of stiffness in virtual environments. *Proceedings of the ASME Dynamic Systems and Control Division.*, 1997.

78. S. Dixon and A. Cooke. Shoe-Surface Interaction in Tennis. *Biomedical engineering principles in sports*, page 125, 2004.

79. K. v. d. Doel. Physically based models for liquid sounds. *ACM Trans. Appl. Percept.*, 2(4):534–546, 2005.

80. L. Dominjon, A. Lecuyer, J. Burkhardt, G. Andrade-Barroso, and S. Richir. The "Bubble" technique: interacting with large virtual environments using haptic devices with limited workspace. In *Proceedings of World Haptics conference*, pages 639–640, 2005.

81. C. Drioli and D. Rocchesso. Acoustic rendering of Particle-Based simulation of liquids in motion. In *Proceedings of the Int. Conference on digital Audio Effects*, Como, Italy, Sept. 2009.

82. P. Dupont, V. Hayward, B. Armstrong, and F. Altpeter. Single state elastoplastic friction models. *IEEE Transactions on Automatic Control*, 47(5):787–792, 2002.

83. H. Durrant-Whyte and T. C. Henderson. Multisensor data fusion. In *Springer Handbook of Robotics*. Springer-Verlag, 2008.

84. A. Ebrahimpour, A. Hamam, R. L. Sack, and W. Patten. Measuring and modeling dynamic loads imposed by moving crowds. *Journal of Structural Engineering*, pages 1468–1474, 1996.

85. A. Ekimov and J. Sabatier. Vibration and sound signatures of human footsteps in buildings. *J. Acoust. Soc. Am.*, page 762, Jan 2006.

86. A. Ekimov and J. Sabatier. A review of human signatures in urban environments using acoustic and seismic methods. In *Proc. of IEEE Technologies for Homeland Security*, 2008.

87. A. Ekimov and J. M. Sabatier. Human motion analyses using footstep ultrasound and doppler ultrasound. *Journal of the Acoustical Society of America*, 2008(6), 123.

88. D. P. W. Ellis. *Prediction-driven computational auditory scene analysis*. PhD thesis, 1996. Supervisor-Barry L. Vercoe.

89. D. Engel, C. Curio, L. Tcheang, B. Mohler, and H. H. BÃŒlthoff. A psychophysically calibrated controller for navigating through large environments in a limited free-walking space. In *Proceedings of the 2008 ACM symposium on Virtual reality software and technology*, pages 157–164, Bordeaux, France, 2008. ACM.

90. A. J. Eronen, V. T. Peltonen, J. T. Tuomi, A. P. Klapuri, S. Fagerlund, T. Sorsa, G. Lorho, and J. Huopaniemi. Audio-based context recognition. *IEEE Transactions on Audio, Speech and Language Processing*, 14(1):321–329, 2006.

91. A. J. Farnell. Marching onwards – procedural synthetic footsteps for video games and animation. In *pd Convention*, 2007.

92. J. P. Fiene and K. J. Kuchenbecker. Shaping Event-Based haptic transients via an improved understanding of real contact dynamics. In *EuroHaptics Conference, 2007 and Symposium on Haptic Interfaces for Virtual Environment and Teleoperator Systems. World Haptics 2007. Second Joint*, pages 170–175. IEEE, 2007.

93. F. Fontana and R. Bresin. Physics-based sound synthesis and control: crushing, walking and running by crumpling sounds. In *Proc. Colloquium on Musical Informatics*, pages 109–114, Florence, Italy, May 2003.

94. F. Fontana, F. Morreale, T. Regia-Corte, A. Lécuyer, and M. Marchal. Auditory recognition of floor surfaces by temporal and spectral cues of walking. In *Proc. Int. Conf. on Auditory Display*, Budapest, Hungary, Jun. 20-24 2011.

95. D. Fox, J. Hightower, H. Kauz, L. Liao, and D. Patterson. Bayesian techniques for location estimation. In *Proceedings of The 2003 Workshop on Location-Aware Computing*, page 16, 2003.

96. D. Fox, J. Hightower, L. Liao, D. Schulz, and G. Borriello. Bayesian filtering for location estimation. *IEEE Pervasive Computing*, 2(3):24–33, 2003.

97. V. Fox, J. Hightower, L. Liao, D. Schulz, and G. Borriello. Bayesian filtering for location estimation. *Pervasive Computing, IEEE*, 2(3):24–33, July-Sept. 2003.

98. G. J. Franz. Splashes as sources of sound in liquids. *The Journal of the Acoustical Society of America*, 31(8):1080–1096, 1959.

99. Y. Freund and R. E. Schapire. Experiments with a new boosting algorithm. In *International Conference on Machine Learning*, pages 148–156, 1996.

100. P. Froehlich, R. Simon, and L. Baille. Mobile spatial interaction special issue. *Personal and Ubiquitous Computing*, (In press), 2008.

101. M. Fukumoto and T. Sugimura. Active click: tactile feedback for touch panels. In *CHI'01 extended abstracts*, pages 121–122. ACM, 2001.

102. T. Funkhouser, N. Tsingos, and J. Jot. Survey of methods for modeling sound propagation in interactive virtua l environment systems. 2003.

103. D. Gafurov, K. Helkala, and T. Sondrol. Biometric gait authentication using accelerometer sensor. *Journal of Computers*, 1(7):51–59, 2006.

104. D. Gafurov, K. Helkala, and T. Sondrol. Gait recognition using acceleration from mems. In *ARES*, pages 432–439, 2006.

105. F. Galbraith and M. Barton. Ground loading from footsteps. *Journal of the Acoustical Society of America*, Jan 1970.

106. W. W. Gaver. How Do We Hear in the World? Explorations in Ecological Acoustics. *Ecological Psychology*, 5(4):285–313, Apr. 1993.

107. W. W. Gaver. What in the world do we hear?: An ecological approach to auditory event perception. *Ecological psychology*, 5(1):1–29, 1993.

108. Z. Ghahramani and G. E. Hinton. Parameter estimation for linear dynamical systems. Technical Report (Short Note) CRG-TR-96-2, Department of Computer Science, University of Toronto, 1996.

109. Z. Ghahramani and G. E. Hinton. Parameter estimation for linear dynamical systems. Technical Report CRG-TR-96-2, University of Toronto, 1996.

110. B. Giordano, Y. Visell, H.-Y. Yao, V. Hayward, J. Cooperstock, and S. McAdams. Audio-haptic identification of ground materials during walking. *J. of the Acoustical Society of America*, 2011.

111. B. L. Giordano, S. McAdams, Y. Visell, J. R. Cooperstock, H. Yao, and V. Hayward. Non-visual identification of walking grounds. In *Proc. of Acoustics'08 in J. Acoust. Soc. Am.*, volume 123 (5), page 3412, 2008.

112. V. Hayward. Physically-based haptic synthesis. In M. Lin and M. Otaduy, editors, *Haptic Rendering: Foundations, Algorithms and Applications*. AK Peters, Ltd, 2007.

113. V. Hayward, O. Astley, M. Cruz-Hernandez, D. Grant, and G. Robles-De-La-Torre. Haptic interfaces and devices. *Sensor Review*, 24(1):16–29, 2004.

114. V. Hayward and K. Maclean. Do it yourself haptics, part I. *IEEE Robotics and Automation*, December 2007.

115. R. Headon and R. Curwen. Recognizing movements from the ground reaction force. In *Proceedings of the Workshop on Perceptive User Interfaces*, 2001.

116. M. Heintz. Real walking in virtual learning environments: Beyond the advantage of naturalness. In U. Cress, V. Dimitrova, and M. Specht, editors, *Learning in the Synergy of Multiple Disciplines*, volume 5794, pages 584–595. Springer Berlin Heidelberg, Berlin, Heidelberg, 2009.

117. E. A. Heinz, K. steven Kunze, S. Sulistyo, H. Junker, P. Lukowicz, and G. Tröster. Experimental evaluation of variations in primary features used for accelerometric context recognition. In *Proc. of the 1st European Symposium on Ambient Intelligence (EUSAI 2003*, pages 252–263, 2003.

118. H. Herrmann and S. Roux. *Statistical models for the fracture of disordered media*. North Holland, 1990.

119. E. Hoffmann. A comparison of hand and foot movement times. *Ergonomics*, 34(4):397, 1991.

120. J. A. Hogan. The past recaptured: Marcel Proust's aesthetic theory. *Ethics*, 49:187–203, January 1939.

121. J. Hollerbach. Locomotion interfaces and rendering. In M. Lin and M. Otaduy, editors, *Haptic Rendering: Foundations, Algorithms and Applications*. A K Peters, Ltd, 2008.

122. J. Hollerbach, D. Checcacci, H. Noma, Y. Yanagida, and N. Tetsutani. Simulating side slopes on locomotion interfaces using torso forces. In *Proceedings of International Symposium on Haptic Interfaces for Virtual Environment and Teleoperator Systems*, page 91, 2003.

123. J. Hollerbach, R. Mills, D. Tristano, R. Christensen, W. Thompson, and Y. Xu. Torso force feedback realistically simulates slope on treadmill-style locomotion interfaces. *International Journal of Robotics Research*, 20(12):939–952, 2001.

124. J. Hollerbach, Y. Xu, R. Christensen, and S. Jacobsen. Design specifications for the second generation sarcos treadport locomotion interface. In *Proceedings of Haptics Symposium*, pages 1293–1298, 2000.

125. S. H. Holzreiter and M. E. Köhle. Assessment of gait patterns using neural networks. *J. Biomech*, 26(6):645–51, 1993.

126. K. Houston and D. McGaffigan. Spectrum analysis techniques for personnel detection using seismic sensors. In *Proceedings of SPIE*, volume 5090, page 162, 2003.

127. K. M. Houston and D. P. McGaffigan. Spectrum analysis techniques for personnel detection using seismic sensors. *Proc. of SPIE*, 5090:162–173, 2003.

128. S. Howison, J. Ockendon, and J. Oliver. Deep- and shallow-water slamming at small and zero deadrise angles. *Journal of Engineering Mathematics*, 42(3):373–388, Apr. 2002.

129. K. H. Hunt and F. R. E. Crossley. Coefficient of restitution interpreted as damping in vibroimpact. *ASME J. Applied Mech.*, pages 440–445, June 1975.

130. T. Huynh and B. Schiele. Analyzing features for activity recognition. In *sOc-EUSAI '05: Proceedings of the 2005 joint conference on Smart objects and ambient intelligence*, pages 159–163, New York, NY, USA, 2005. ACM.

131. A. J. Ijspeert, J. Nakanishi, and S. Schaal. Learning attractor landscapes for learning motor primitives. In *Advances in Neural Information Processing Systems 15*, pages 1547–1554. MIT Press, 2003.

132. J. A. Ijspeert, J. Nakanishi, and S. Schaal. Learning rhythmic movements by demonstration using nonlinear oscillators. In *IEEE International Conference on Intelligent Robots and Systems (iros 2002)*, pages 958–963, 2002.

133. V. Interrante, B. Ries, and L. Anderson. Seven league boots: A new metaphor for augmented locomotion through moderately large scale immersive virtual environments. In *Proceedings of the IEEE Symposium on 3D User Interfaces*, pages 167—170, 2007.

134. A. Itai and H. Yasukawa. Personal identification using footstep detection based on wavelets. In *Proc. of ISPACS*, 2006.

135. A. Itai and H. Yasukawa. Footstep classification using wavelet decomposition. *Communications and Information Technologies, 2007. ISCIT '07. International Symposium on*, pages 551–556, Oct. 2007.

136. A. Itai and H. Yasukawa. Footstep classification using simple speech recognition technique. *Circuits and Systems, 2008. ISCAS 2008. IEEE International Symposium on*, pages 3234–3237, May 2008.

137. H. Iwata. Walking about virtual environments on an infinite floor. In *Proc. of IEEE Virtual Reality*, 1999.

138. H. Iwata, H. Yano, and F. Nakaizumi. Gait master: A versatile locomotion interface for uneven virtual terrain. In *Proceedings of IEEE Virtual Reality Conference*, pages 131–137, 2001.

139. H. Iwata, H. Yano, and H. Tomioka. Powered shoes. In *Proceedings of SIGGRAPH 2006 Emerging technologies*, page 28, 2006.

140. J. Jacko and A. Sears. *The human-computer interaction handbook: fundamentals, evolving technologies, and emerging applications.* Lawrence Erlbaum Assoc Inc, 2003.

141. V. Jousmaki and R. Hari. Parchment-skin illusion: sound-biased touch. *Current Biology*, 8(6):R190–R191, 1998.

142. R. Kalman and R. Bucy. New results in linear filtering and prediction theory. *ASME Transactions Part D J. Basic Engrg.*, pages 95–108, 1961.

143. M. Kaltenbrunner, T. Bovermann, R. Bencina, and E. Costanza. Tuio: A protocol for tabletop tangible user interfaces. In *Proc of Gesture Workshop 2005*. Gesture Workshop, 2005.

144. J. Kekoni, H. Hämäläinen, J. Rautio, and T. Tukeva. Mechanical sensibility of the sole of the foot determined with vibratory stimuli of varying frequency. *Experimental brain research*, 78(2):419–424, 1989.

145. N. Kern and B. Schiele. Context-aware notification for wearable computing. In *ISWC*, pages 223–230. IEEE Computer Society, 2003.

146. N. Kern, B. Schiele, and A. Schmidt. Recognizing context for annotating a live life recording. *Personal and Ubiquitous Computing*, 11(4):251–263, 2007.

147. R. L. Klatzky, D. K. Pai, and E. P. Krotkov. Perception of material from contact sounds. *Presence: Teleoperators and Virtual Environment*, 9(4):399–410, 2000.

148. J. F. Knight, H. W. Bristow, S. Anastopoulou, C. Baber, A. Schwirtz, and T. N. Arvanitis. Uses of accelerometer data collected from a wearable system. *Personal Ubiquitous Comput.*, 11(2):117–132, 2007.

149. L. Kohli, E. Burns, D. Miller, and H. Fuchs. Combining passive haptics with redirected walking. In *Proceedings of the 2005 international conference on Augmented tele-existence*, pages 253–254, Christchurch, New Zealand, 2005. ACM.

150. P. Korpipää, M. Koskinen, J. Peltola, S.-M. Mäkelä, and T. Seppänen. Bayesian approach to sensor-based context awareness. *Personal and Ubiquitous Computing*, 7(2):113–124, 2003.

151. A. Krause, D. P. Siewiorek, A. Smailagic, and J. Farringdon. Unsupervised, dynamic identification of physiological and activity context in wearable computing. In *ISWC*, pages 88–97. IEEE Computer Society, 2003.

152. W. L. Kuan and C. Y. San. Constructivist physics learning in an immersive, multi-user hot air balloon simulation program (iHABS). In *ACM SIGGRAPH 2003 Educators Program*, page 1. ACM Press, 2003.

153. K. J. Kuchenbecker, J. Fiene, and G. Niemeyer. Improving contact realism through event-based haptic feedback. *IEEE Transactions on Visualization and Computer Graphics*, 12(2):219–230, 2006.

154. K. Kunze, P. Lukowicz, H. Junker, and G. Tröster. Where am i: Recognizing on-body positions of wearable sensors. In *In: LOCA'04: International Workshop on Locationand Context-Awareness*, pages 264–275. Springer-Verlag, 2005.

155. K. V. Laerhoven and O. Cakmakci. What shall we teach our pants? In *ISWC '00: Proceedings of the 4th IEEE International Symposium on Wearable Computers*, page 77, Washington, DC, USA, 2000. IEEE Computer Society.

156. K. V. Laerhoven and H.-W. Gellersen. Spine versus porcupine: A study in distributed wearable activity recognition. In *ISWC '04: Proceedings of the Eighth International Symposium on Wearable Computers*, pages 142–149, Washington, DC, USA, 2004. IEEE Computer Society.

157. J. J. LaViola, D. A. Feliz, D. F. Keefe, and R. C. Zeleznik. Hands-free multi-scale navigation in virtual environments. In *Proceedings of the ACM symposium on Interactive 3D graphics*, pages 9–15. ACM, 2001.

158. J. LaViola Jr, D. Feliz, D. Keefe, and R. Zeleznik. Hands-free multi-scale navigation in virtual environments. In *Proc of the 2001 symposium on Interactive 3D graphics*, pages 9–15. ACM New York, NY, USA, 2001.

159. A. Law, B. Peck, Y. Visell, P. Kry, and J. Cooperstock. A multi-modal floor-space for displaying material deformation underfoot in virtual reality. In *Proc. of the IEEE Intl. Workshop on Haptic Audio Visual Environments and Their Applications*, 2008.

160. A. Lécuyer. Simulating haptic feedback using vision: a survey of research and applications of pseudo-haptic feedback. *Presence: Teleoperators and Virtual Environments*, 18(1):39–53, 2009.

161. A. Lécuyer, J.-M. Burkhardt, and L. Etienne. Feeling bumps and holes without a haptic interface: the perception of pseudo-haptic textures. In *Proceedings of SIGCHI Conference on Human factors in computing systems*, pages 239–246, 2004.

162. A. Lécuyer, J.-M. Burkhardt, J.-M. Henaff, and S. Donikian. Camera motions improve sensation of walking in virtual environments. In *Proceedings of IEEE International Conference on Virtual Reality*, pages 11–18, 2006.

163. A. Lécuyer, J.-M. Burkhardt, J.-M. Henaff, and S. Donikian. Camera motions improve the sensation of walking in virtual environments. In *Proceedings of IEEE Virtual Reality*, pages 11–18, 2006.

164. S. Lederman. Auditory texture perception. *Perception*, 1979.

165. M.-Y. Lee, K.-S. Soon, and C.-F. Lin. New computer protocol with subsensory stimulation and visual/auditory biofeedback for balance assessment in amputees. *J. of Computers*, 4(10):1005–1011, Oct. 2009.

166. S.-W. Lee and K. Mase. Activity and location recognition using wearable sensors. *IEEE Pervasive Computing*, 1(3):24–32, 2002.

167. E. A. Lehmann. *Particle Filtering Methods for Acoustic Source Localisation and Tracking*. PhD thesis, Australian National University, 2004.

168. M. Lesser. Thirty years of liquid impact research: a tutorial review. *Wear*, 186-187(Part 1):28–34, July 1995.

169. J. Lester, T. Choudhury, N. Kern, G. Borriello, and B. Hannaford. A hybrid discriminative/generative approach for modeling human activities. In L. P. Kaelbling and A. Saffiotti, editors, *IJCAI-05, Proceedings of the Nineteenth International Joint Conference on Artificial Intelligence, Edinburgh, Scotland, UK, July 30-August 5, 2005*, pages 766–772. Professional Book Center, 2005.

170. X. Li, R. J. Logan, and R. E. Pastore. Perception of acoustic source characteristics: Walking sounds. *Journal of the Acoustical Society of America*, 90(6):3036–3049, 1991.

171. L. Liao. *Location-based activity recognition*. PhD thesis, University of Washington, 1996.

172. L. Liao, D. Fox, and H. Kautz. Extracting places and activities from gps traces using hierarchical conditional random fields. *Int. J. Robotics Research*, 2007.

173. L. Liao, D. J. Patterson, D. Fox, and H. Kautz. Learning and inferring transportation routines. *Artificial Intelligence*, 171(5–6):311–331, 2007.

174. M. S. Longuet-Higgins. An analytic model of sound production by raindrops. *Journal of Fluid Mechanics*, 214:395–410, 1990.

175. P. Lukowicz, J. A. Ward, H. Junker, M. Stäger, G. Tröster, A. Atrash, and T. Starner. Recognizing workshop activity using body worn microphones and accelerometers. In A. Ferscha and F. Mattern, editors, *Pervasive Computing, Second International Conference, PERVASIVE 2004, Vienna, Austria, April 21-23, 2004, Proceedings*, volume 3001 of *Lecture Notes in Computer Science*, pages 18–32. Springer, 2004.

176. P. Lukowicz, J. A. Ward, H. Junker, M. Stäger, G. Tröster, A. Atrash, and T. Starner. Recognizing workshop activity using body worn microphones and accelerometers. In A. Ferscha and F. Mattern, editors, *Pervasive Computing, Second International Conference, PERVASIVE 2004, Vienna, Austria, April 21-23, 2004, Proceedings*, volume 3001 of *Lecture Notes in Computer Science*, pages 18–32. Springer, 2004.

177. A. MacEachren, G. Cai, R. Sharma, I. Rauschert, I. Brewer, L. Bolelli, B. Shaparenko, S. Fuhrmann, and H. Wang. Enabling collaborative geoinformation access and decision-making through a natural, multimodal interface. *International Journal of Geographical Information Science*, 19(3):293–317, 2005.

178. I. MacKenzie. Fitts' law as a research and design tool in human-computer interaction. *Human-Computer Interaction*, 7(1):91–139, 1992.

179. M. Mahvash and V. Hayward. Haptic rendering of cutting: A fracture mechanics approach. *Haptics-e*, 2(3):1–12, 2001.

180. J. Mantyjarvi, M. Lindholm, E. Vildjiounaite, S.-M. Makela, and H. J. Ailisto. Identifying users of portable devices from gait pattern with accelerometers. In *Proc. IEEE Intl. Conf. on Acoustics, Speech, and Signal Processing*, 2005.

181. M. J. Mathie, A. C. F. Coster, N. H. Lovell, and B. G. Celler. TOPICAL REVIEW: Accelerometry: providing an integrated, practical method for long-term, ambulatory monitoring of human movement. *Physiological Measurement*, 25:1–+, Apr. 2004.

182. G. P. Mazarakis and J. N. Avaritsiotis. A prototype sensor node for footstep detection. In *Wireless Sensor Networks, 2005. Proceeedings of the Second European Workshop on*, pages 415–418, 2005.

183. T. McGuine and J. Keene. The effect of a balance training program on the risk of ankle sprains in high school athletes. *The American journal of sports medicine*, 34(7):1103, 2006.

184. R. Meir and G. Rätsch. An introduction to boosting and leveraging. pages 118–183, 2003.

185. L. Middleton, A. Buss, A. Bazin, and M. Nixon. A floor sensor system for gait recognition. In *Proceedings of the Fourth IEEE Workshop on Automatic Identification Advanced Technologies*, pages 171–176, 2005.

186. T. Minka. Bayesian inference in dynamical models – an overview. Web published, Aug. 2008. `http://research.microsoft.com/users/minka/papers/dynamic.html`.

187. M. Minnaert. On musical air-bubbles and the sounds of running water. *Philosophical Magazine Series 7*, 16(104):235 – 248, 1933.

188. D. Minnen, T. Starner, J. Ward, P. Lukowicz, and G. Troster. Recognizing and discovering human actions from on-body sensor data. *Multimedia and Expo, IEEE International Conference on*, 0:1545–1548, 2005.

189. D. Mitrovic. Discrimination and retrieval of environmental sounds. Master's thesis, Vienna University of Technology, 2005.

190. T. Moeslund and E. Granum. A survey of computer vision-based human motion capture. *Computer Vision and Image Understanding*, 81(3):231–268, 2001.

191. T. Moeslund, A. Hilton, and V. Krueger. A survey of advances in vision-based human motion capture and analysis. *Computer vision and image understanding*, 104(2-3):90–126, 2006.

192. J. J. Monaghan. Smoothed particle hydrodynamics. *Annual Review of Astronomy and Astrophysics*, 30(1):543–574, Sept. 1992.

193. M. Morioka, D. Whitehouse, and M. Griffin. Vibrotactile thresholds at the fingertip, volar forearm, large toe, and heel. *Somatosensory & Motor Research*, 25(2):101–112, 2008.

194. H. Morishita, R. Fukui, and T. Sato. High resolution pressure sensor distributed floor for future human-robot symbiosis environments. In *IEEE/RSJ International Conference on Intelligent Robots and Systems (IROS 2002), Lausanne, Switzerland*, pages 1246–1251, 2002.

195. W. Moss, H. Yeh, J. Hong, M. C. Lin, and D. Manocha. Sounding liquids: Automatic sound synthesis from fluid simulation. *ACM Trans. Graph.*, 29(3):1–13, 2010.

196. A. Mostayed, M. M. G. Mazumder, S. Kim, and S. J. Park. Abnormal gait detection using discrete fourier transform. In *MUE*, pages 36–40, 2008.

197. M. Muller, D. Charypar, and M. Gross. Particle-based fluid simulation for interactive applications. In *Proceedings of the 2003 ACM SIGGRAPH/Eurographics symposium on Computer animation*, pages 154–159, San Diego, California, 2003. Eurographics Association.

198. M. Muller, B. Solenthaler, R. Keiser, and M. Gross. Particle-based fluid-fluid interaction. In *Proceedings of the 2005 ACM SIGGRAPH/Eurographics symposium on Computer animation*, pages 237–244, Los Angeles, California, 2005. ACM.

199. T. Murakita, T. Ikeda, and H. Ishiguro. Human tracking using floor sensors based on the Markov chain Monte Carlo method. In *Pattern Recognition, 2004. ICPR 2004. Proceedings of the 17th International Conference on*, volume 4, 2004.

200. K. Murphy. *Dynamic Bayesian Network : Representation, Inference and Learning*. PhD thesis, The University of California at Berkeley, 2002.

201. R. Murray-Smith, J. Williamson, S. Hughes, and T. Quaade. Stane: synthesized surfaces for tactile input. In M. Czerwinski, A. M. Lund, and D. S. Tan, editors, *Proceedings of the 2008 Conference on Human Factors in Computing Systems, CHI 2008, 2008, Florence, Italy, April 5-10, 2008*, pages 1299–1302. ACM, 2008.

202. A. Nashel and S. Razzaque. Tactile virtual buttons for mobile devices. In *Proceedings of CHI*, pages 854–855. ACM, 2003.

203. S. Nasuno, A. Kudrolli, A. Bak, and J. Gollub. Time-resolved studies of stick-slip friction in sheared granular layers. *Physical Review E*, 58(2):2161–2171, 1998.

204. S. Nasuno, A. Kudrolli, and J. Gollub. Friction in granular layers: hysteresis and precursors. *Physical Review Letters*, 79(5):949–952, 1997.

205. N. Nitzsche, U. D. Hanebeck, and G. Schmidt. Motion compression for telepresent walking in large target environments. *Presence: Teleoper. Virtual Environ.*, 13(1):44–60, 2004.

206. H. Noma. Design for locomotion interface in a large-scale virtual environment. atlas: Atr locomotion interface for active self-motion. In *Proceedings of the ASME Dynamic Systems and control division*, pages 111–118, 1998.

207. H. Noma, T. Sugihara, and T. Miyasato. Development of ground surface simulator for tel-e-merge system. In *Proceedings of IEEE Virtual Reality Conference*, page 217, 2000.

208. R. Nordahl. Increasing the motion of users in photorealistic virtual environments by utilizing auditory rendering of the environment and ego-motion. *Proceedings of Presence*, pages 57–62, 2006.

209. R. Nordahl, A. Berrezag, S. Dimitrov, L. Turchet, V. Hayward, and S. Serafin. Preliminary experiment combining virtual reality haptic shoes and audio synthesis. *Haptics: Generating and Perceiving Tangible Sensations*, pages 123–129, 2010.

210. R. Nordahl, S. Serafin, and L. Turchet. Sound Synthesis and Evaluation of Interactive Footsteps for Virtual Re ality Applications. In *Proc. IEEE Virtual Reality*, 2010.

211. D. Norman. *The Design of Future Things*. Basic Books, New York, 2007.

212. B. North, A. Blake, M. Isard, and J. Rittscher. Learning and classification of complex dynamics. *IEEE Transactions on Pattern Analysis and Machine Intelligence*, 22:1016–1034, 2000.

213. J. F. O'Brien and J. K. Hodgins. Graphical modeling and animation of brittle fracture. In *Proceedings of the 26th annual conference on Computer graphics and interactive techniques*, SIGGRAPH '99, pages 137–146, New York, NY, USA, 1999. ACM Press/Addison-Wesley Publishing Co.

214. A. M. Okamura, M. R. Cutkosky, and J. T. Dennerlein. Reality-based models for vibration feedback in virtual environments. *IEEE/ASME Transactions on Mechatronics*, 6(3):245–252, 2001.

215. A. M. Okamura, J. T. Dennerlein, and R. D. Howe. Vibration feedback models for virtual environments. In *1998 IEEE International Conference on Robotics and Automation, 1998. Proceedings*, volume 1, pages 674–679 vol.1. IEEE, 1998.

216. R. J. Orr and G. D. Abowd. The smart floor: a mechanism for natural user identification and tracking. In *ACM CHI extended abstracts*, pages 275–276, New York, NY, USA, 2000. ACM.

217. N. Ouarti, A. Lécuyer, and A. Berthoz. Procede de simulation de mouvements propres par retour haptique et dispositif mettant en oeuvre le procede. French Patent no. 09 56406, Sept. 2009. Filed on Sep. 17, 2009.

218. D. Pai, K. Doel, D. James, J. Lang, J. Lloyd, J. Richmond, and S. Yau. Scanning physical interaction behavior of 3d objects. In *Proceedings of the 28th annual conference on Computer graphics and interactive techniques*, pages 87–96, 2001.

219. A. Pakhomov and T. Goldburt. Seismic signals and and noise assesment for footstep detection range estimation in different environments. In *Proc. of SPIE*, volume 5417, 2004.

220. A. Pakhomov, A. Sicignano, M. Sandy, and T. Goldburt. Seismic footstep signal characterization. In *Proc. of SPIE*, volume 5071, pages 297–305, 2003.

221. T. Pakkanen and R. Raisamo. Appropriateness of foot interaction for non-accurate spatial tasks. In *CHI '04: CHI '04 extended abstracts on Human factors in computing systems*, pages 1123–1126, New York, NY, USA, 2004. ACM.

222. M. Palaniswami and R. Begg. *Computational Intelligence for Movement Sciences: Neural Networks and Other Emerging Techniques (Computational Intelligence and Its Applications Series)*. IGI Publishing, Hershey, PA, USA, 2006.

223. S. Papetti, F. Avanzini, and D. Rocchesso. Energy and accuracy issues in numerical simulations of a non-linear impact model. In *Proc. Conf. on Digital Audio Effects (DAFX-09)*, Como, Italy, September 2009.

224. S. Papetti, M. Civolani, F. Fontana, A. Berrezag, and V. Hayward. Audio-tactile display of ground properties using interactive shoes. In *Proc. 5th Int. Haptic and Auditory Interaction Design Workshop*, pages 117–128, Sept. 2010.

225. S. Papetti, F. Fontana, and M. Civolani. A shoe-based interface for ecological ground augmentation. In *Proc. 4th Int. Haptic and Auditory Interaction Design Workshop*, volume 2, Dresden, Germany, Sep. 10-11 2009.

226. S. Papetti, F. Fontana, and M. Civolani. Effects of ecological auditory and vibrotactile underfoot feedback on human gait: a preliminary investigation. In *Proc. 6th Int. Workshop on Haptic and Audio Interaction Design (HAID 2011)*, Kyoto, Japan, 2011.

227. J. Paradiso, C. Abler, K.-y. Hsiao, and M. Reynolds. The magic carpet: physical sensing for immersive environments. In *ACM CHI extended abstracts*, pages 277–278, New York, NY, USA, 1997. ACM.

228. J. Parkka, M. Ermes, P. Korpipää, J. Mäntyjärvi, J. Peltola, and I. Korhonen. Activity classification using realistic data from wearable sensors. *IEEE Transactions on Information Technology in Biomedicine*, 10(1):119–128, 2006.

229. R. D. Patterson, K. Robinson, J. Holdsworth, D. McKeown, C.Zhang, and M. H. Allerhand. Complex sounds and auditory images. In Y. Cazals, L. Demany, and K. Horner, editors, *Auditory Physiology and Perception*. Pergamon, 1992.

230. M. Pauly, R. Keiser, B. Adams, P. Dutré, M. Gross, and L. J. Guibas. Meshless animation of fracturing solids. In *ACM Transactions on Graphics (TOG)*, SIGGRAPH '05, New York, NY, USA, 2005. ACM. ACM ID: 1073296.

231. R. Pausch, T. Burnette, D. Brockway, and M. E. Weiblen. Navigation and locomotion in virtual worlds via flight into hand-held miniatures. In *Proceedings of ACM SIGGRAPH*, pages 399–400. ACM, 1995.

232. G. Pearson and M. Weiser. Of moles and men: the design of foot controls for workstations. In *Proc of the ACM SIGCHI conf on Human factors in Comp Sys*, pages 333–339. ACM New York, NY, USA, 1986.

233. T. Peck, M. Whitton, and H. Fuchs. Evaluation of reorientation techniques for walking in large virtual environments. In *Virtual Reality Conference, 2008. VR '08. IEEE*, pages 121–127, 2008.

234. L. Peltola, C. Erkut, P. R. Cook, and V. Välimäki. Synthesis of hand clapping sounds. *IEEE Trans. Audio, Speech and Language Proc.*, 15(3):1021–1029, 2007.

235. V. Peltonen, J. Tuomi, A. Klapuri, J. Huopaniemi, and T. Sorsa. Computational auditory scene recognition. In *Proc. of IEEE Conf. on Acoustics, Speech and Signal Processing (ICASSP'02)*, 2002.

236. S. Perreault and C. Gosselin. Cable-driven parallel mechanisms: Application to a locomotion interface. *Journal of Mechanical Design*, 130:102301, 2008.

237. J. Perry. *Gait analysis: normal and pathological function*. SLACK incorporated, 1992.

238. R. F. Pinkston. A touch sensitive dance floor/midi controller. *Journal of the Acoustical Society of America*, 96(5):3302–3302, 1994.

239. S. Pirttikangas, J. Suutala, J. Riekki, and J. Röning. Learning vector quantization in footstep identification. In *Proceedings of 3rd IASTED International Conference on Artificial Intelligence and Applications (AIA), IASTED*, 2003.

240. A. Polaine. The flow principle in interactivity. In *Proceedings of the 2nd Au. conf. on Interactive Entertainment*, page 158, 2005.

241. I. Potamitis, H. Chen, and G. Tremoulis. Tracking of multiple moving speakers with multiple microphone arrays. *Speech and Audio Processing, IEEE Transactions on*, 12(5):520–529, Sept. 2004.

242. R. L. Potter, L. J. Weldon, and B. Shneiderman. Improving the accuracy of touch screens: an experimental evaluation of three strategies. In *CHI '88: Proc of the SIGCHI conference on Human factors in computing systems*, pages 27–32, New York, NY, USA, 1988. ACM.

243. I. Poupyrev, S. Maruyama, and J. Rekimoto. Ambient touch: designing tactile interfaces for handheld devices. In *Proceedings of UIST*. ACM, 2002.

244. A. Quarteroni, R. Sacco, and F. Saleri. *Numerical Mathematics*. Springer, 2nd edition, 2007.

245. L. R. Rabiner and B. H. Juang. An introduction to hidden Markov models. *IEEE ASSP Magazine*, 3(1):4–16, Jan 1986.

246. R. Radhakrishnan and D. Divakaran. Generative process tracking for audio analysis. In *IEEE International Conference on Acoustics, Speech and Signal Processing*, 2006.

247. A. Rahimi. *Learning to transform time Series with a Few Examples*. PhD thesis, Massachussetts Institute of Technology, 2006.

248. C. Ramstein and V. Hayward. The pantograph: A large workspace haptic device for a multimodal human-computer interaction. In *Proceedings of the SIGCHI conference on Human factors in computing systems, CHI'04, ACM/SIGCHI Companion-4/94*, pages 57–58, 1994.

249. C. Randell and H. Muller. Context awareness by analyzing accelerometer data. In *ISWC '00: Proceedings of the 4th IEEE International Symposium on Wearable Computers*, page 175, Washington, DC, USA, 2000. IEEE Computer Society.

250. I. Rauschert, P. Agrawal, R. Sharma, S. Fuhrmann, I. Brewer, and A. MacEachren. Designing a human-centered, multimodal gis interface to support emergency management. In *GIS '02: Proceedings of the 10th ACM international symposium on Advances in geographic information systems*, pages 119–124, New York, NY, USA, 2002. ACM.

251. N. Ravi, N. Dandekar, P. Mysore, and M. L. Littman. Activity recognition from accelerometer data. *American Association for Artificial Intelligence*, 2005.

252. S. Razzaque, Z. Kohn, and M. Whitton. Redirected walking. In *Proc of EUROGRAPHICS*, pages 289–294, 2001.

253. S. Razzaque, Z. Kohn, and M. C. Whitton. Redirected walking. In *Proceedings of Eurographics*, 2001.

254. S. Razzaque, D. Swapp, M. Slater, M. C. Whitton, and A. Steed. Redirected walking in place. In *Proceedings of the workshop on Virtual environments 2002*, pages 123–130, Barcelona, Spain, 2002. Eurographics Association.

255. E. G. Richardson. The sounds of impact of a solid on a liquid surface. *Proceedings of the Physical Society. Section B*, 68(8):541–547, Aug. 1955.

256. I. Rius, X. Varona, F. X. Roca, and J. Gonzàlez. Posture constraints for bayesian human motion tracking. In *Articulated Motion and Deformable Objects*, pages 414–423. Springer Verlag, 2006.

257. D. Rocchesso and F. Fontana, editors. *The Sounding Object*. Edizioni di Mondo Estremo, Florence, Italy, 2003.

258. N. Roman and D. Wang. Binaural tracking of multiple moving sources. *Acoustics, Speech, and Signal Processing, 2003. Proceedings. (ICASSP '03). 2003 IEEE International Conference on*, 5:V–149–52 vol.5, April 2003.

259. S. Roweis and Z. Ghahramani. A unifying review of linear gaussian models. *Neural Computation*, 11(2):305–345, 1999.

260. S. Roweis and Z. Ghahramani. An em algorithm for identification of nonlinear dynamical systems. In *Proc. of Neural Information Processing Systems 13 (NIPS'00)*, 2000.

261. R. A. Ruddle and S. Lessels. The benefits of using a walking interface to navigate virtual environments. *ACM Trans. Comput.-Hum. Interact.*, 16(1):1–18, 2009.

262. K. Salisbury, F. Conti, and F. Barbagli. Haptic rendering: introductory concepts. *IEEE Computer Graphics and Applications*, 24(2):24– 32, 2004.

263. R. Sanders and M. Scorgie. The Effect of Sound Delivery Methods on a User's Sense of Presence in a Virtual Environment, 2002.

264. J. Schacher and M. Neukom. Ambisonics Spatialization Tools for Max/MSP. In *Proceedings of the International Computer Music Conference*, 2006.

265. R. M. Schafer. *The tuning of the world*. Random House Inc., 1977.

266. A. Schmidt, M. Strohbach, K. v. Laerhoven, A. Friday, and H.-W. Gellersen. Context acquisition based on load sensing. In *UbiComp '02: Proceedings of the 4th international conference on Ubiquitous Computing*, pages 333–350, London, UK, 2002. Springer-Verlag.

267. S. Serafin, L. Turchet, R. Nordahl, S. Dimitrov, A. Berrezag, and V. Hayward. Identification of virtual grounds using virtual reality haptic shoes and sound synthesis. In *Proceedings of Eurohaptics symposium on Haptic and Audio-Visual Stimuli: Enhancing Experiences and Interaction*, 2010.

268. J. P. Sethna, K. A. Dahmen, and C. R. Myers. Crackling noise. *Nature*, (410):242–250, Mar. 2001.

269. S. Shimojo and L. Shams. Sensory modalities are not separate modalities: plasticity and interactions. *Current Opinion in Neurobiology*, 11(4):505–509, 2001.

270. Y. Shoji, A. Itai, and H. Yasukawa. Personal identification using footstep detection in indoor environments. *IEICE Trans. Fundamentals*, E88-A(8), 2005.

271. R. H. Shumway and D. S. Stoffer. An approach to time series smoothing and forecasting using the em algorithm. *J. of Time Series Analysis*, 3(4), 1982.

272. J. Sinclair, P. Hingston, and M. Masek. Considerations for the design of exergames. In *Proc of the 5th Int ACM Conf on Comp Graphics and Interactive Techniques, Southeast Asia*, page 295, 2007.

273. M. Slater, M. Usoh, and A. Steed. Taking steps: The influence of a walking technique on presence in virtual reality. *ACM Trans. on Computer-Human Interaction*, 2(3):201–219, 1995.

274. J. Sreng, F. Bergez, J. Legarrec, A. Lécuyer, and C. Andriot. Using an event-based approach to improve the multimodal rendering of 6dof virtual contact. In *Proceedings of ACM Symposium on Virtual Reality Software and Technology (ACM VRST)*, pages 173–179, 2007.

275. M. A. Srinivasan and C. Basdogan. Haptics in virtual environments: Taxonomy, research status, and challenges. *Computers & Graphics*, 21(4):393–404, 1997.

276. M. Stäger, P. Lukowicz, N. Perera, T. von Büren, G. Tröster, and T. Starner. Soundbutton: Design of a low power wearable audio classification system. In *ISWC*, pages 12–17. IEEE Computer Society, 2003.

277. F. Steinicke, G. Bruder, J. Jerald, H. Frenz, and M. Lapp. Estimation of detection thresholds for redirected walking techniques. *IEEE Transactions on Visualization and Computer Graphics*, 16(1):17–27, 2009.

278. F. Steinicke, G. Bruder, T. Ropinski, and K. H. Hinrichs. Moving towards generally applicable redirected walking. In *Proceedings of the Virtual Reality International Conference (VRIC)*, pages 15—24. IEEE Press, 2008.

279. V. Stiles, I. James, S. Dixon, and I. Guisasola. Natural Turf Surfaces: The Case for Continued Research. *Sports Medicine*, 39(1):65, 2009.

280. R. L. Storms and M. J. Zyda. Interactions in Perceived Quality of Auditory-Visual Displays. *Presence: Teleoperators & Virtual Environments*, 9(6):557–580, 2000.

281. J. Su. Motion compression for telepresence locomotion. *Presence: Teleoper. Virtual Environ.*, 16(4):385–398, 2007.

282. A. Subramanya, A. Raj, J. Bilmes, and D. Fox. Recognizing activities and spatial context using wearable sensors. In *In Proc. of the Conference on Uncertainty in Artificial Intelligence (UAI*, 2006.

283. G. Succi, D. Clapp, R. Gampert, and G. Prado. Footstep detection and tracking. In *Proc. of SPIE*, 2001.

284. E. A. Suma, S. Clark, D. Krum, S. Finkelstein, M. Bolas, and Z. Warte. Leveraging change blindness for redirection in virtual environments. In *2011 IEEE Virtual Reality Conference (VR)*, pages 159–166. IEEE, 2011.

285. J. Suutala, K. Fujinami, and J. Roening. Gaussian Process Person Identifier Based on Simple Floor Sensors. In *Proceedings of the 3rd European Conference on Smart Sensing and Context*, page 68. Springer-Verlag, 2008.

286. J. Suutala and J. Röning. Combining classifiers with different footstep feature sets and multiple samples for person identification. In *IEEE International Conference on Acoustics, Speech, and Signal Processing (ICASSP)*, volume 5, 2005.

287. J. Suutala and J. Roning. Methods for person identification on a pressure-sensitive floor: Experiments with multiple classifiers and reject option. *Information Fusion*, 9(1), 2008.

288. M. Tanaka and H. Inoue. A study on walk recognition by frequency analysis of footsteps. *Trans. IEE of Japan*, 119-C(6), 1999.

289. E. M. Tapia, S. Intille, and K. Larson. Real-time recognition of physical activities and their intensities using wireless accelerometers and a heart rate monitor. In *Proceedings of the 11th International Conference on Wearable Computers (ISWC '07)*, 2007.

290. E. M. Tapia, S. S. Intille, and K. Larson. Activity recognition in the home using simple and ubiquitous sensors. pages 158–175. 2004.

291. J. Taylor and Q. Huang. *CRC handbook of electrical filters*. CRC, 1997.

292. J. Templeman, P. Denbrook, and L. Sibert. Virtual locomotion: Walking in place through virtual environments. *Presence*, 8(6):598–617, 1999.

293. J. N. Templeman, P. S. Denbrook, and L. E. Sibert. Virtual locomotion: Walking in place through virtual environments. *Presence: Teleoper. Virtual Environ.*, 8(6):598–617, 1999.

294. L. Terziman, A. Lécuyer, S. Hillaire, and J. M. Wiener. Can camera motions improve the perception of traveled distance in virtual environments? In *Proceedings of IEEE Virtual Reality Conference*, pages 131–134, 2009.

295. S. Thrun, W. Burgard, and D. Fox. *Probabilistic Robotics*. MIT Press, 2005.

296. S. Thrun, W. Burgard, and D. Fox. *Probabilistic Robotics (Intelligent Robotics and Autonomous Agents)*. The MIT Press, September 2005.

297. H. Tramberend, F. Hasenbrink, G. Eckel, U. Lechner, and M. Goebel. CyberStage – An Advanced Virtual Environment. *ERCIM News*, 31, October 1997.

298. L. Turchet, M. Marchal, A. Lécuyer, R. Nordahl, and S. Serafin. Influence of auditory and visual feedback for perceiving walking over bumps and holes in desktop vr. In *Proceedings of the 17th ACM Symposium on Virtual Reality Software and Technology*, pages 139–142. ACM, 2010.

299. L. Turchet, M. Marchal, A. Lécuyer, S. Serafin, and R. Nordahl. Influence of visual feedback for perceiving walking over bumps and holes in desktop vr. In *Proceedings of 17th ACM Symposium on Virtual Reality Software and Technology*, pages 139–142, 2010.

300. L. Turchet and S. Serafin. A preliminary study on sound delivery methods for footstep sounds. In *Proc. Conf. on Digital Audio Effects (DAFX-11)*, Paris, France, Sept. 2011.

301. L. Turchet, S. Serafin, S. Dimitrov, and R. Nordahl. Conflicting audio-haptic feedback in physically based simulation of walking sounds. *Haptic and Audio Interaction Design*, pages 97–106, 2010.

302. L. Turchet, S. Serafin, and R. Nordahl. Examining the role of context in the recognition of walking sounds. In *Proceedings of Sound and Music Computing Conference*, 2010.

303. L. Turchet, S. Serafin, and R. Nordahl. Physically based sound synthesis and control of footsteps sounds. In *Proc. Conf. on Digital Audio Effects (DAFX-10)*, Graz, Austria, Sep. 6-10 2010.

304. M. Usoh, K. Arthur, M. C. Whitton, R. Bastos, A. Steed, M. Slater, and J. Frederick P. Brooks. Walking > walking-in-place > flying, in virtual environments. In *Proceedings of the 26th annual conference on Computer graphics and interactive techniques*, pages 359–364. ACM Press/Addison-Wesley Publishing Co., 1999.

305. K. van den Doel, P. Kry, and D. Pai. FoleyAutomatic: physically-based sound effects for interactive simulation and animation. *Proceedings of the 28th annual conference on Computer graphics and interactive techniques*, pages 537–544, 2001.

306. Y. Visell and J. Cooperstock. Enabling gestural interaction by means of tracking dynamical systems models and assistive feedback. In *Proc. of the IEEE Intl. Conf. on Systems, Man, and Cybernetics*, Montreal, 2007.

307. Y. Visell and J. Cooperstock. Design of a Vibrotactile Display via a Rigid Surface. In *Proc. of IEEE Haptics Symposium*, 2010.

308. Y. Visell, J. Cooperstock, B. L. Giordano, K. Franinovic, A. Law, S. McAdams, K. Jathal, and F. Fontana. A vibrotactile device for display of virtual ground materials in walking. In *Proc. of Eurohaptics 2008*, 2008.

309. Y. Visell, F. Fontana, B. Giordano, R. Nordahl, S. Serafin, and R. Bresin. Sound design and perception in walking interactions. *International Journal of Human-Computer Studies*, 2009.

310. Y. Visell, F. Fontana, B. Giordano, R. Nordahl, S. Serafin, and R. Bresin. Sound design and perception in walking interactions. *Int. J. Human-Computer Studies*, (67):947–959, 2009.

311. Y. Visell, B. Giordano, G. Millet, and J. Cooperstock. Vibration influences haptic perception of surface compliance during walking. *PloS one*, 6(3):e17697, 2011.

312. Y. Visell, A. Law, and J. Cooperstock. Touch Is Everywhere: Floor Surfaces as Ambient Haptic Interfaces. *IEEE Transactions on Haptics*, 2009.

313. Y. Visell, A. Law, J. Ip, S. Smith, and J. R. Cooperstock. Interaction Capture in Immersive Virtual Environments via an Intelligent Floor Surface. In *Proc. of IEEE Virtual Reality*, 2010 (To appear).

314. Y. Visell, A. Law, S. Smith, R. Rajalingham, and J. Cooperstock. Contact sensing and interaction techniques for a distributed multimodal floor display. In *Proc. of IEEE 3DUI*, 2010.

315. Y. Visell, S. Smith, A. Law, R. Rajalingham, and J. Cooperstock. Contact sensing and interaction techniques for a distributed, multimodal floor display. In *3D User Interfaces (3DUI), 2010 IEEE Symposium on*, pages 75–78. IEEE, 2010.

316. E. Wan and R. van der Merwe. *The Unscented Kalman Filter*. Wiley Publishing, 2001.

317. E. A. Wan and R. V. D. Merwe. The unscented kalman filter for nonlinear estimation. *Adaptive Systems for Signal Processing, Communications, and Control Symposium 2000. AS-SPCC. The IEEE 2000*, pages 153–158, 2000.

318. J. Wang. Gaussian process dynamical models for human motion. *IEEE Trans. on Pattern Analysis and Machine Intelligence*, 30(2), 2008.

319. J. Wang, D. Fleet, and A. Hertzmann. Gaussian process dynamical models. In *Advances in Neural Information Processing Systems 18*, 2005.

320. J. A. Ward. *Activity monitoring: continuous recognition and performance evaluation*. PhD thesis, ETH Zürich, Switzerland, 2006.

321. J. A. Ward, P. Lukowicz, G. Troster, and T. E. Starner. Activity recognition of assembly tasks using body-worn microphones and accelerometers. *IEEE Trans. Pattern Analysis and Machine Intelligence*, 28(10):1553–1567, Oct. 2006.

322. B. G. Watters. Impact noise characteristics of female hard-heeled foot traffic. *J. Acoust. Soc. Am.*, 37:619–630, 1965.

323. L. Wauben, M. van Veelen, D. Gossot, and R. Goossens. Application of ergonomic guidelines during minimally invasive surgery: a questionnaire survey of 284 surgeons. *Surgical endoscopy*, 20(8):1268–1274, 2006.

324. E. Weinstein, K. Steele, A. Agarwal, and J. Glass. Loud: A 1020-node modular microphone array and beamformer for intelligent computing spaces. Technical report, Massachusetts Institute of Technology, 2004.

325. A. Westner and V. M. B. Jr. Applying blind source separation and deconvolution to real-world acoustic environments. In *Proc. of the AES*, 1999.

326. A. Westner and V. M. B. Jr. Blind separation of real world audio signals using overdetermined mixtures. In *Proc. Int. Conf. on Independent Component Analysis and Blind Source Separation*, pages 251–256, 1999.

327. B. Williams, G. Narasimham, B. Rump, T. P. McNamara, T. H. Carr, J. Rieser, and B. Bodenheimer. Exploring large virtual environments with an HMD when physical space is limited. In *Proceedings of the ACM symposium on Applied perception in graphics and visualization*, pages 41–48, Tubingen, Germany, 2007. ACM.

328. X. Xie, Q. Lin, H. Wu, G. Narasimham, T. P. McNamara, J. Rieser, and B. Bodenheimer. A system for exploring large virtual environments that combines scaled translational gain and interventions. In *Proceedings of the 7th symposium on Applied perception in graphics and visualization*, page 65. ACM Press, 2010.

329. H.-F. Xing, F. Li, and Y.-L. Liu. Wavelet denoising and feature extraction of seismic signal for footstep detection. *Wavelet Analysis and Pattern Recognition, 2007. ICWAPR '07. International Conference on*, 1:218–223, Nov. 2007.

330. T. Yamada, S. Nakamura, and K. Shikano. Robust speech recognition with speaker localization by a microphone array. In *Proc. ICSLP '96*, volume 3, pages 1317–1320, Philadelphia, PA, Oct. 1996.

331. K. Yin and D. K. Pai. Footsee: an interactive animation system. In *SCA '03: Proceedings of the 2003 ACM SIGGRAPH/Eurographics symposium on Computer animation*, pages 329–338, Aire-la-Ville, Switzerland, Switzerland, 2003. Eurographics Association.

332. C. A. Zanbaka, B. C. Lok, S. V. Babu, A. C. Ulinski, and L. F. Hodges. Comparison of path visualizations and cognitive measures relative to travel technique in a virtual environment. *IEEE Transactions on Visualization and Computer Graphics*, 11(6):694–705, 2005.

333. S. Zhai. *Human performance in six degree of freedom input control*. PhD thesis, University of Toronto, 1995.

334. Y. Zhang, J. Pettré, Q. Peng, and S. Donikian. Data based steering of virtual human using a velocity-space approach. In *Proceedings of Motion in Games*, Lecture Notes in Computer Science 5884, pages 170–181. Springer, 2009.